Roundabouts, Truck Safety, Motorcycles, and Mopeds 2018

Transportation Research Record: Journal of the Transportation Research Board Volume 2672, Issue 34

ISBN: 978-1-5108-8243-0

Printed from e-media with permission by:

Curran Associates, Inc.
57 Morehouse Lane
Red Hook, NY 12571

Some format issues inherent in the e-media version may also appear in this print version.

Published by Sage Publications USA

ISBN (Print) 978-1-5108-8243-0
ISBN 2018 Printed Set (All Issues) 978-1-5108-7735-1

Printed by Curran Associates, Inc. (2019)

Additional copies of this publication are available from:

Curran Associates, Inc.
57 Morehouse Lane
Red Hook, NY 12571 USA
Phone: 845-758-0400
Fax: 845-758-2633
Email: curran@proceedings.com
Web: www.proceedings.com

Volume 2672 Number 34 December 2018

TRR
JOURNAL OF THE TRANSPORTATION RESEARCH BOARD

Contents

Roundabouts, Truck Safety, Motorcycles, and Mopeds 2018

2018 FRED BURGGRAF AWARD: Safety and System Users

Articles

2018 PATRICIA F. WALLER AWARD: Outstanding Paper on Safety and System Users

Articles

Transportation Officials, Washington, D.C.
K. Jane Williams, *Acting Administrator, Federal Transit Administration, U.S. Department of Transportation*
Paul Zukunft *(Admiral, U.S. Coast Guard), Commandant, U.S. Coast Guard, Washington, D.C.*

---

* Membership as of February 2019.

TRB

TRANSPORTATION RESEARCH BOARD

CORE SPONSORS

Transportation Departments of the 50 States and the District of Columbia

Federal Government

U.S. Department of Transportation
- Federal Aviation Administration (FAA)
- Federal Highway Administration
- Federal Motor Carrier Safety Administration (FMCSA)
- Federal Railroad Administration (FRA)
- Federal Transit Administration
- National Highway Traffic Safety Administration (NHTSA)
- Office of the Assistant Secretary of Transportation for Research and Technology (OST-R)

U.S. Air Force Civil Engineer Center (AFCEC)
U.S. Coast Guard (USCG)
U.S. Department of Energy
U.S. Department of the Interior (DOI)
U.S. Environmental Protection Agency (EPA)

Nongovernmental Organizations

American Public Transportation Association (APTA)
Association of American Railroads (AAR)
California Air Resources Board

The National Academies of
SCIENCES · ENGINEERING · MEDICINE

The **National Academy of Sciences** was established in 1863 by an Act of Congress, signed by President Lincoln, as a private, nongovernmental institution to advise the nation on issues related to science and technology. Members are elected by their peers for outstanding contributions to research. Dr. Marcia McNutt is president.

The **National Academy of Engineering** was established in 1964 under the charter of the National Academy of Sciences to bring the practices of engineering to advising the nation. Members are elected by their peers for extraordinary contributions to engineering. Dr. C. D. Mote, Jr., is president.

The **National Academy of Medicine** (formerly the Institute of Medicine) was established in 1970 under the charter of the National Academy of Sciences to advise the nation on medical and health issues. Members are elected by their peers for distinguished contributions to medicine and health. Dr. Victor J. Dzau is president.

The three Academies work together as the National Academies of Sciences, Engineering, and Medicine to provide independent, objective analysis and advice to the nation and conduct other activities to solve complex problems and inform public policy decisions. The Academies also encourage education and research, recognize outstanding contributions to knowledge, and increase public understanding in matters of science, engineering, and medicine.

Learn more about the National Academies of Sciences, Engineering, and Medicine at **www.national-academies.org.**

The **Transportation Research Board** is one of seven major programs of the National Academies of Sciences, Engineering, and Medicine. The mission of the Transportation Research Board is to increase the benefits that transportation contributes to society by providing leadership in transportation innovation and progress through research and information exchange, conducted within a setting that is objective, interdisciplinary, and multimodal. The Board's varied activities annually engage about 7,000 engineers, scientists, and other transportation researchers and practitioners from the public and private sectors and academia, all of whom contribute their expertise in the public interest. The program is supported by state transportation departments, federal agencies including the component administrations of the U.S. Department of Transportation, and other organizations and individuals interested in the development of transportation.

Learn more about the Transportation Research Board at **www.TRB.org**.

TRR
JOURNAL OF THE TRANSPORTATION RESEARCH BOARD

Transportation Research Record
2018, Vol. 2672(34) 1–9

DOI: 10.1177/0361198118770174
journals.sagepub.com/home/trr
SAGE

Utilization of Gaps at Single-Lane Roundabouts

Nathan P. Belz[1] and Addison Yang[2]

Abstract

The study of gap acceptance behavior at roundabout entries is important for both capacity and safety analysis. Although considerable research exists on the relative size of vehicular gaps that are accepted during the driver decision process when entering a roundabout, the effective use of accepted gaps has not been adequately addressed. In the same way as follow-up time between entering vehicles is deemed to be a critical component in the entry capacity of a roundabout and considered in gap-acceptance derived capacity models, it is proposed here that the temporal distribution associated with when drivers enter into usable gaps is of equal importance. Approximately 60 hours of video were collected at six single-lane roundabouts located in Maine, New Hampshire, Vermont, New York, and Alaska, from which the gap decision processes of drivers were extracted and analyzed. A total of 1,181 accepted gaps were observed under saturated conditions and were included in the final dataset. The data show statistically significant temporal variations in the utilization of usable gaps by drivers, as well as between-site differences across the study locations. These findings suggest potential merit in the consideration of this factor in the calibration and capacity analysis process for single-lane roundabouts.

Roundabouts are a particularly advantageous form of traffic control as they allow for near-continuous movement of vehicles into and through the intersection. Inherently, this flow is directly related to the nature of a drivers' decision of whether or not to enter the roundabout based on other drivers already circulating on the roundabout; a behavioral process typically referred to as *gap acceptance*. The study of gap acceptance behavior at roundabout entries is important for both capacity and safety. Although considerable research exists on the relative size of vehicular gaps that are accepted during the driver decision process when entering a roundabout, the effective use of accepted gaps has not been adequately addressed.

To elaborate, when a driver enters into a gap in the circulating traffic stream on a roundabout there is some amount of space, less than the full headway, between two circulating vehicles (i.e., a leading and lagging pair of circulating vehicles) which is considered as the usable gap (illustrated as t_u in Figure 1). The term "gap" corresponds to the area being utilized by an entering driver and is confined to the physical space between the two circulating vehicles. In practice, headways are used as a proxy for gaps because of the relative ease of in-field collection and inherently includes a small portion of time related to the length of the leading vehicle. Presumably, there will be more usable time on the leading end of a gap closer to the lead vehicle (V_1) in the circulating pair than there will be on the lagging end of the gap closer to the lagging vehicle (V_2) in the circulating pair. It holds that there will be a certain point in time up to which a driver who enters into that usable gap will have no influence on the lagging vehicle of the gap pair. A driver entering after this point will impede the lagging vehicle and consequently diminish the capacity of the roundabout. Much like the critical gap and follow-up time between entering vehicles are deemed to be critical components of the entry capacity of a roundabout and are considered in the gap-acceptance derived capacity model presented in the 2010 Highway Capacity Manual (HCM2010), it is proposed here that the temporal distribution of where drivers enter into a usable gap is of equal importance.

The overall research question being addressed here is; how are drivers utilizing the gaps that they accepted at single-lane roundabout approaches? Using field data, this study focuses on a single component of the driving task at roundabouts and seeks to make a distinction between the nature of simply accepting a gap and effectively utilizing a gap. A framework was developed using real-world data to isolate gap acceptance events and then characterize those events based on the way in which those gaps were used.

The objectives of this paper are threefold. First, to propose the concept of gap utilization and how it can be defined using timestamped turning movement counts. In essence, how is gap utilization different than gap acceptance? Second,

[1]Department of Civil and Environmental Engineering, University of Alaska Fairbanks, Fairbanks, AK
[2]Alaska Department of Transportation and Public Facilities, Northern Region, Fairbanks, AK

Corresponding Author:
Address correspondence to Nathan P. Belz: npbelz@alaska.edu

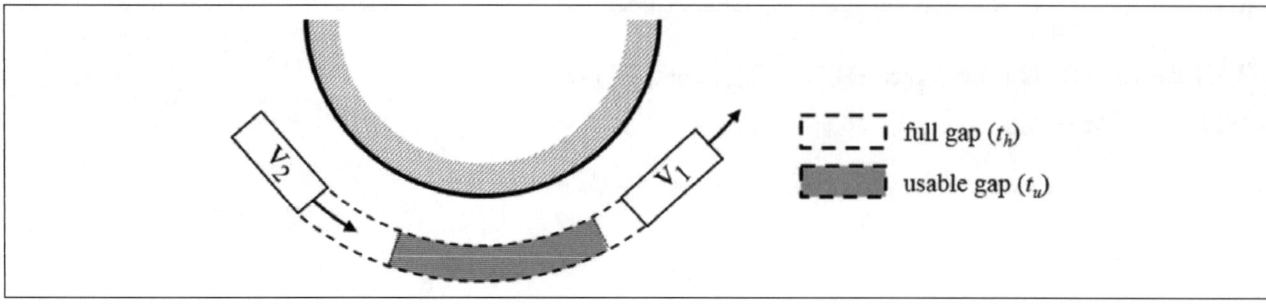

Figure 1. An example of a full gap (t_h) and usable gap (t_u).

determine if gap utilization varies between roundabout locations and relative to the age of the roundabout. Practically, can gap utilization serve to account for between-site variations in capacity estimates unaccounted for in other modeling approaches? Finally, examine the differences in gap utilization between priority taking and non-priority taking type entry behavior. Can gap utilization account for entry behavior types that impede the circulating flow of traffic (e.g., priority taking and priority reversal events). To this end, approximately 60 hours of video were collected at six single-lane roundabouts located in Maine, New Hampshire, Vermont, New York, and Alaska from which the gap decision processes of drivers were extracted and analyzed.

Background

Driver behavior is arguably the largest contributory factor in the performance of a roundabout. This is reflected in the most recent release of the HCM 2010 (*1*), which provided an update to the calibration process based on local driver behavior and changes in driver experience over time through the use of critical gap and follow-up time. Although there have been significant efforts to expand the knowledge of roundabout operations in the United States, there is still a need for comprehensive data collection and analysis efforts related to driver behavior.

Of all the driving tasks that a driver is presented with at a roundabout, the gap acceptance process is by far the most complex and has the most influence on the capacity and safety of the intersection. Existing gap acceptance frameworks are grounded in theories originally developed for stop-controlled intersections and assume that a driver who is waiting on a minor approach will only enter into a gap in the major stream of traffic if a safe opportunity (i.e., a gap of large enough size) is presented (*2, 3*). Aggressive driver behavior is then characterized by either a single critical gap or a distribution of critical gaps that serve as a proxy for variations in driver aggressiveness. Previous studies indicate that the typically accepted gap opportunities range from four to five seconds at roundabout approaches (*4, 5*).

Other work has recognized the limitations and stochastic nature of gap-acceptance behavior, yet these approaches tend to be more computationally intensive and still rely on the basic assumption that all drivers are consistent both in time and with respect to each other (*6–9*). Furthermore, the improvement from representing "true stochasticity" with more rigorous approaches (*10, 11*) may only be marginal at best (*12*). Additionally, inconsistencies in driver behavior and variations in traffic conditions make it inappropriate to consider a critical gap value as being generally applicable.

One major limitation of the HCM 2010 gap-acceptance method is the failure to account for the way in which entering vehicles may impact circulating vehicles based on their judgment of the gap. Critical gap is limited in that it cannot account for inconsistencies in driver behavior and variations in traffic conditions (*13–15*). However, the HCM 2010 does indicate that simplified gap-acceptance models may not produce reliable results when there are effective changes in priority. This is also supported by research indicating that merging and priority reversal phenomena observed at roundabouts are reasons for not using gap-acceptance methodologies (*4, 16*). As priority rules at roundabouts (i.e., yield-at-entry) are less strict than at other intersection types, one would expect there to be more variability in the way in which drivers utilize gaps. In this paper, it is proposed that by accounting for the way in which an entering driver utilizes a gap (not just the size of the gap itself), some of the shortcomings of the gap-acceptance methods can be addressed.

This research builds on previous work that addressed types of drivers who deliberately ignored, or were unfamiliar with, intended yield-at-entry traffic rules at roundabouts (*17, 18*). As a roundabout only has a finite amount of space in which to handle perturbations in traffic, the effective use of gaps (particularly at high circulating flows) is critical. Although some research indicates that driver variability and "rule breaking" has the potential to increase capacity at roundabouts, it is improbable that a significant number of drivers entering into gaps on the lagging end would have a net positive impact on overall roundabout capacity (*19, 20*). More likely, this type of behavior would disrupt traffic and cause significant delays to circulating vehicles. One may contend that a driver who enters into the leading end of gap (of an appropriate size) will not be of detriment to roundabout capacity, whereas a driver who enters into the lagging end of a gap (of the same size) is likely to cause a disruption

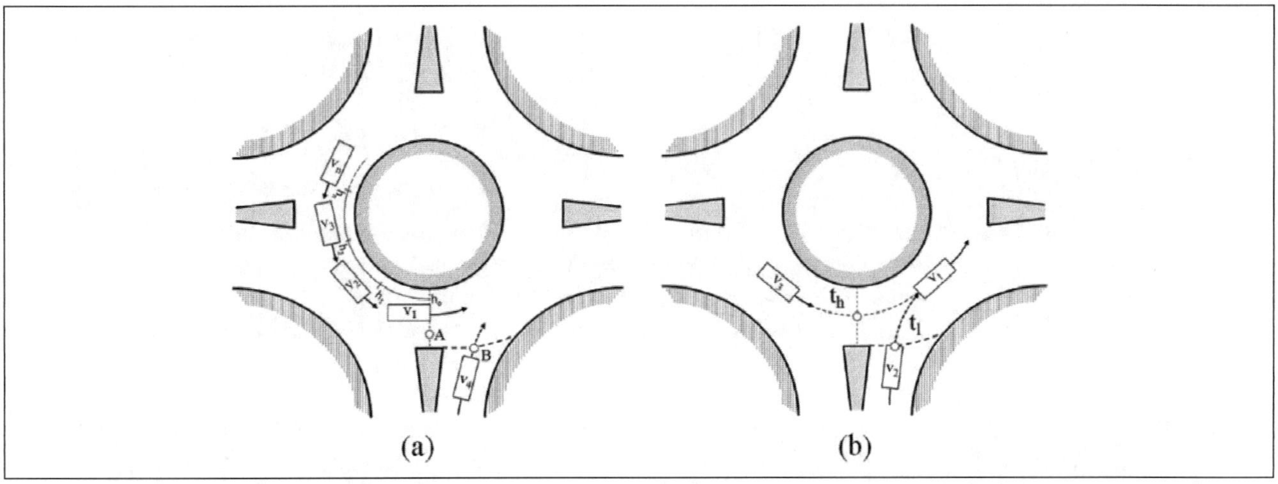

Figure 2. Reference points for (*a*) turning movement timestamp extraction and (*b*) lead portion and total length of accepted gap.

Table 1. Characteristics and Descriptive Statistics of Observed Single-Lane Roundabouts

Location	Year constructed	Recordings (hh:mm)	Accepted gaps extracted	Average accepted gap size (sec)	Average utilization factor
Montpelier, VT	1995	10:07	165	5.16	0.375
Middlebury, VT	2010	09:54	199	4.09	0.400
Glens Falls, NY	2007	09:50	648	4.25	0.585
New London, NH	2009	09:45	35	5.10	0.529
Bangor, ME	2007	09:52	51	4.34	0.551
North Pole, AK	2007	10:15	83	6.78	0.307

to some degree. Furthermore, late entry into a usable gap may serve as a marker for certain types of driver behavior phenomena at roundabouts, such as priority sharing (*21*), limited priority mode (*22*), and priority taking (*17, 23*). To this end, the research presented here explores the concept of gap utilization and proposes its suitability for inclusion as a calibration parameter in gap-acceptance based methods.

Data and Methods

Previously obtained video data obtained by Belz et al. and additionally collected data in North Pole, AK totaling 60 hours of observations were reduced to obtain vehicle headways in the circulatory roadway and the time associated with when a vehicle on the approach entered into the circulatory roadway (see Figure 2) (*17*). The videos were collected on one approach at each of six single-lane roundabouts in Alaska, Vermont, New Hampshire, Maine, and New York (see Table 1 and Figure 3). Each roundabout approach was selected so that there would be similar geometries, to control for as much between-site geometric variation as possible. In particular, the analyzed entry was selected so that the angle and distance to the next upstream approach would be as close to 90° as possible and would be comparable across all sites. This is a critical component of the analysis as prior research indicates that roundabouts designed with low conflict angles are more likely to promote merging-like behavior and thereby facilitate priority reversal when the circulating flows are high (*12, 22*).

Videos were only collected on fair weather days when pavement conditions were dry. The camera was oriented so that the points associated with entering and circulating movements were visible. Video processing was completed with the aid of a timestamp-based keystroke command script to mark the circulating (i.e., headways) and entering vehicle events. Only true accepted gaps were extracted from the raw data (i.e., the entering vehicle event had to be bounded by two circulating vehicle events to be valid). These gaps must have occurred when the approach was under saturated conditions (i.e., two or more vehicles were queued at the time the entry maneuver was initiated). Inherently, these events correspond only to those in which the "gap" was accepted by a single vehicle. Furthermore, only events including passenger vehicles (i.e., both the entering vehicle and the two bounding vehicles) were retained for final analysis.

The timestamp corresponding with an entering vehicle event represents the initiation of an entering maneuver (i.e., crossing the point corresponding to the boundary of the

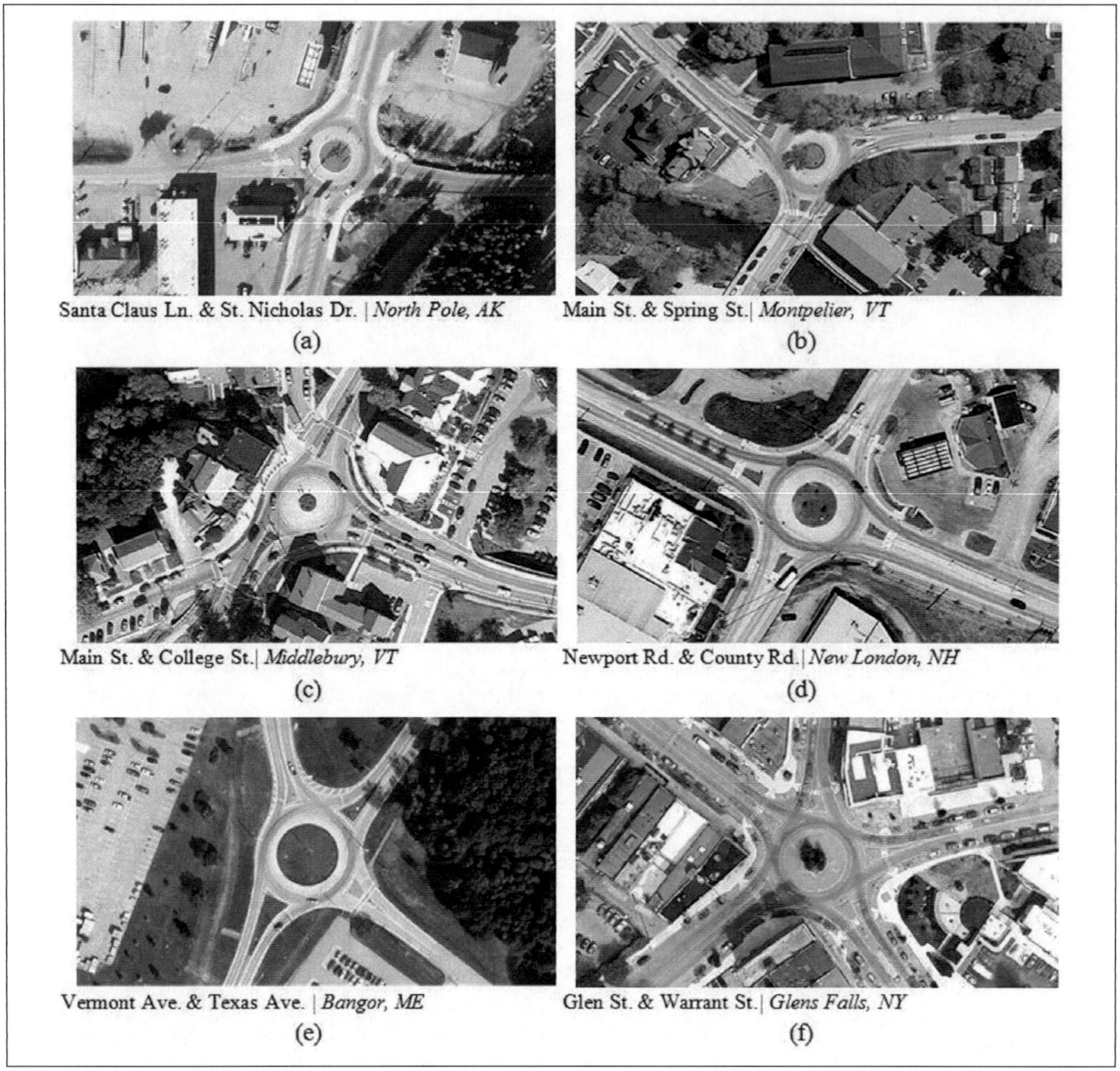

Figure 3. Aerial view of observed roundabouts in (*a*) Alaska, (*b* and *c*) Vermont, (*d*) New Hampshire, (e) Maine, and (*f*) New York.

inscribed circle diameter; point B in Figure 2*a*) and is followed by successful entry into the gap. This timestamp is used as a proxy for the point within the accepted gap that the driver entered. The short amount of time required to traverse from the point of entry to the midpoint of the lane in the circulating stream is assumed to be negligible. Having a timestamp associated with each circulating vehicle (taken at Point A shown in Figure 2*a*), as well as the time a vehicle entered the roundabout allows for the accepted gap to be broken in a lead portion and a lag portion and thus define the nature of how that gap was utilized. In addition, each entering event was flagged if it resulted in a priority taking event so that a comparison could be made between those that directly impeded the circulating flow of traffic and those that did not (*17*).

The lead portion of the accepted "gap" (t_{li}) is then divided by the full length of the accepted "gap" (t_{hi}) for each entering vehicle to compute the gap utilization ratio (g_{ui}) as shown in Figure 2*b*. The gap utilization ratio is subtracted from a value of one so that the gap utilization factor t_g (Equation 1) is represented in terms of the remaining portion of the gap before the arrival of the lagging vehicle in the circulating pair. The

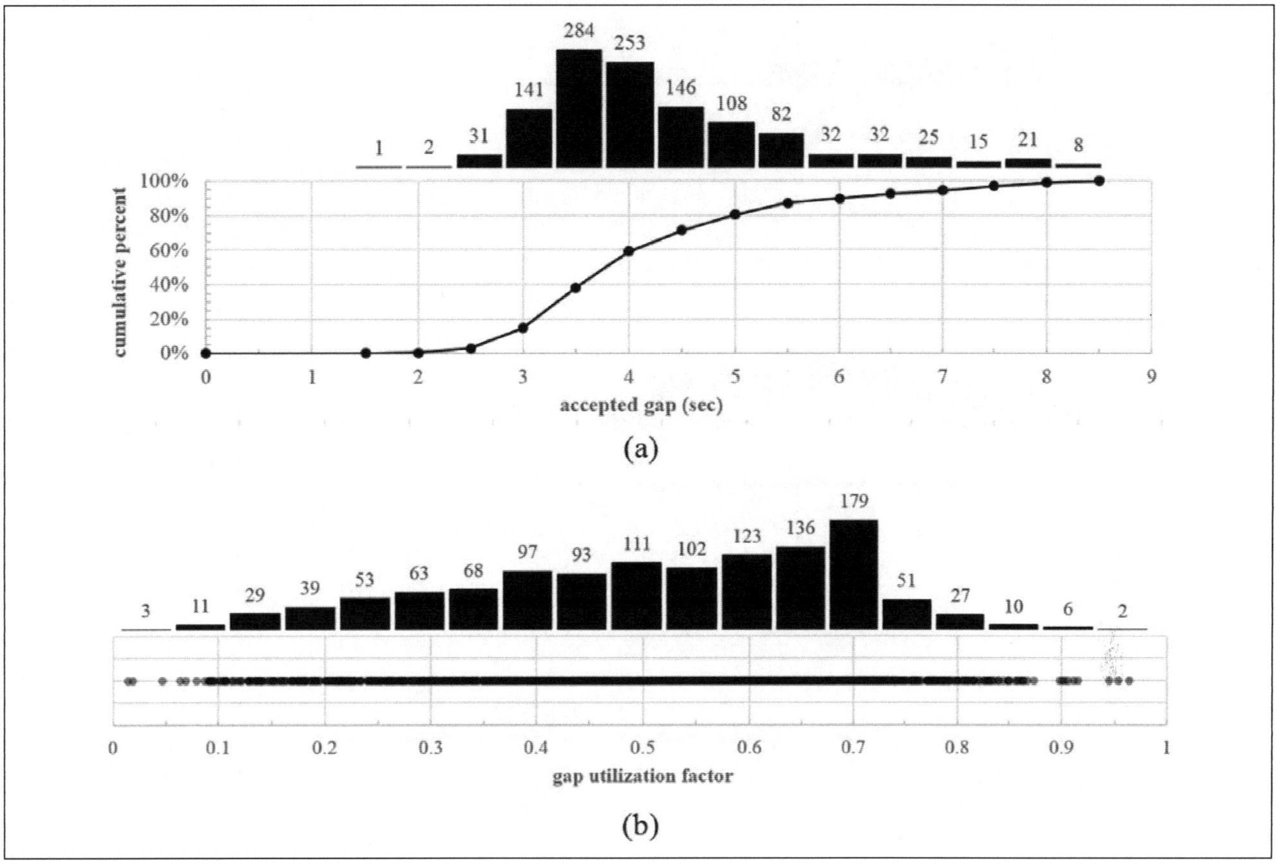

Figure 4. Distribution of (*a*) accepted gaps and (*b*) gap utilization factor for all study sites.

average gap size and the average gap utilization factor for each roundabout are presented in Table 1.

$$t_g = \frac{1}{n}\sum_{i=1}^{n} 1 - g_{ui} = \frac{1}{n}\sum_{i=1}^{n} 1 - \frac{t_{li}}{t_{hi}} \tag{1}$$

where:

t_g = gap utilization factor;

g_{ui} = gap utilization ratio of vehicle i;

t_{li} = lead portion of the accepted gap (headway, in seconds);

t_{hi} = total length of the accepted gap (headway, in seconds).

Results and Analysis

A total of 1,181 accepted gaps under saturated conditions were observed (see Table 1) with an overall average accepted gap of 4.48 seconds and an average gap utilization factor of 0.499. Figure 4 shows the overall accepted gap and gap utilization factor distributions. Figure 5 shows the gap utilization factor distributions for each state.

New York and Alaska have the highest and lowest average gap utilization factor with values of 0.585 and 0.307, respectively. Figure 5 also shows a noticeable shift in the gap utilization factor distributions between each state. Accepted gap size and lead portion of the gap were both plotted against the gap utilization factor (Figures 6 and 7, respectively) for each state. No trend is seen with relation to accepted gap size, but a significant relationship can be seen between lead portion of the gap and the gap utilization factor.

In all cases, the logarithmic function provided a statistically significant improvement in the overall fit as compared with the linear function (Figure 7). Alaska, Vermont, and New Hampshire all have reasonable model fits with r-squared values equal to or exceeding 0.69. Maine and New York show a marginal fit with r-squared values of 0.49 and 0.53, respectively. The nature of this relationship fits with the constraint that a utilization ratio cannot exceed a value of one. Furthermore, these relationships show that with the exception of a few cases, the gap utilization ratio does not exceed a value of 0.8, meaning that roughly 20% of the leading end of the gap between a circulating pair is not used. With the exception of Alaska (Figure 7*a*) and Vermont (Figure 7*b*), the lower limit of the utilization ratio is about 0.2, indicating that roughly 20% of the lagging end of the gap between a circulating pair is not used. This is opposite to the original

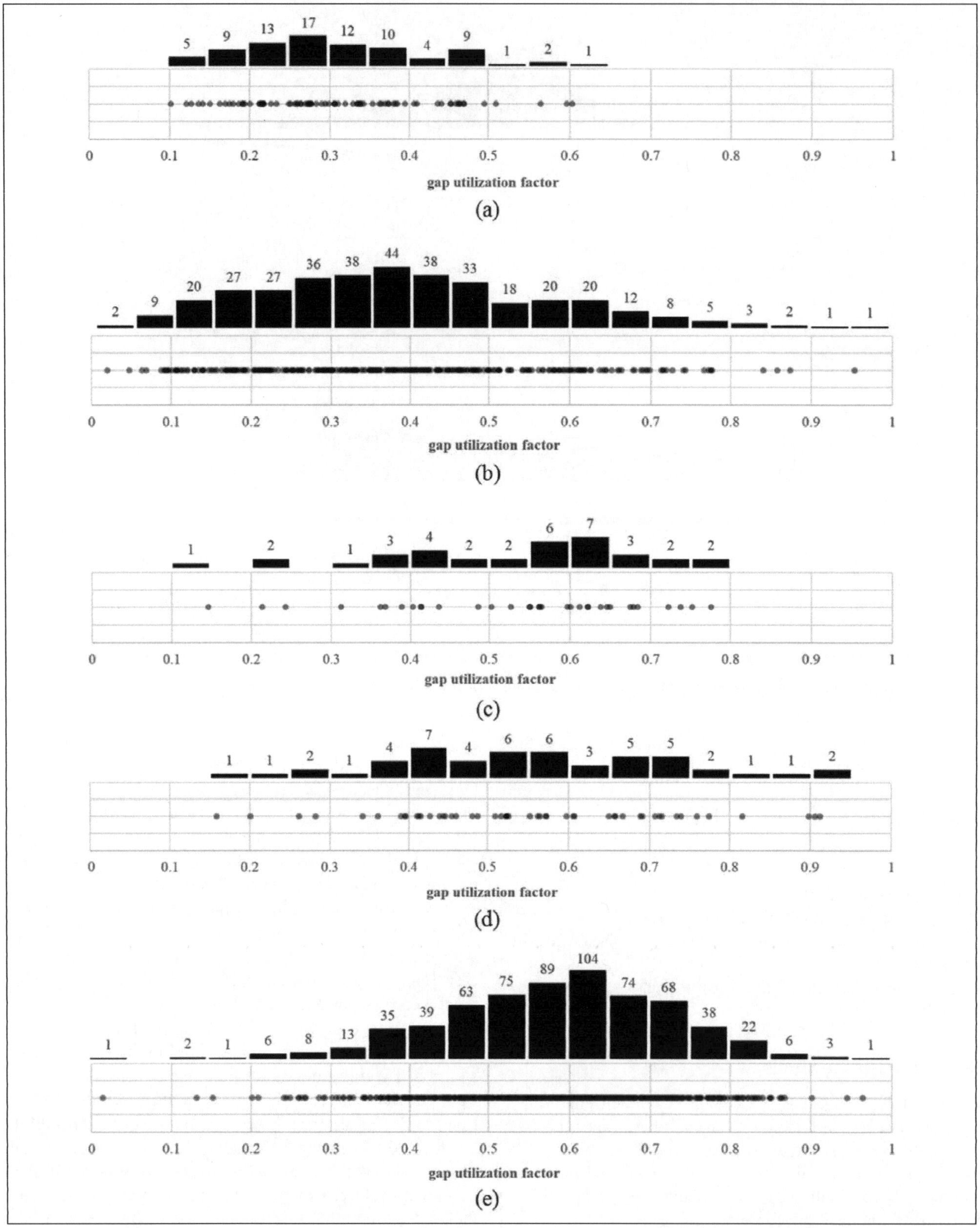

Figure 5. Distribution of gap utilization factor at the (*a*) Alaska, (*b*) Vermont, (*c*) New Hampshire, (*d*) Maine, and (*e*) New York roundabouts.

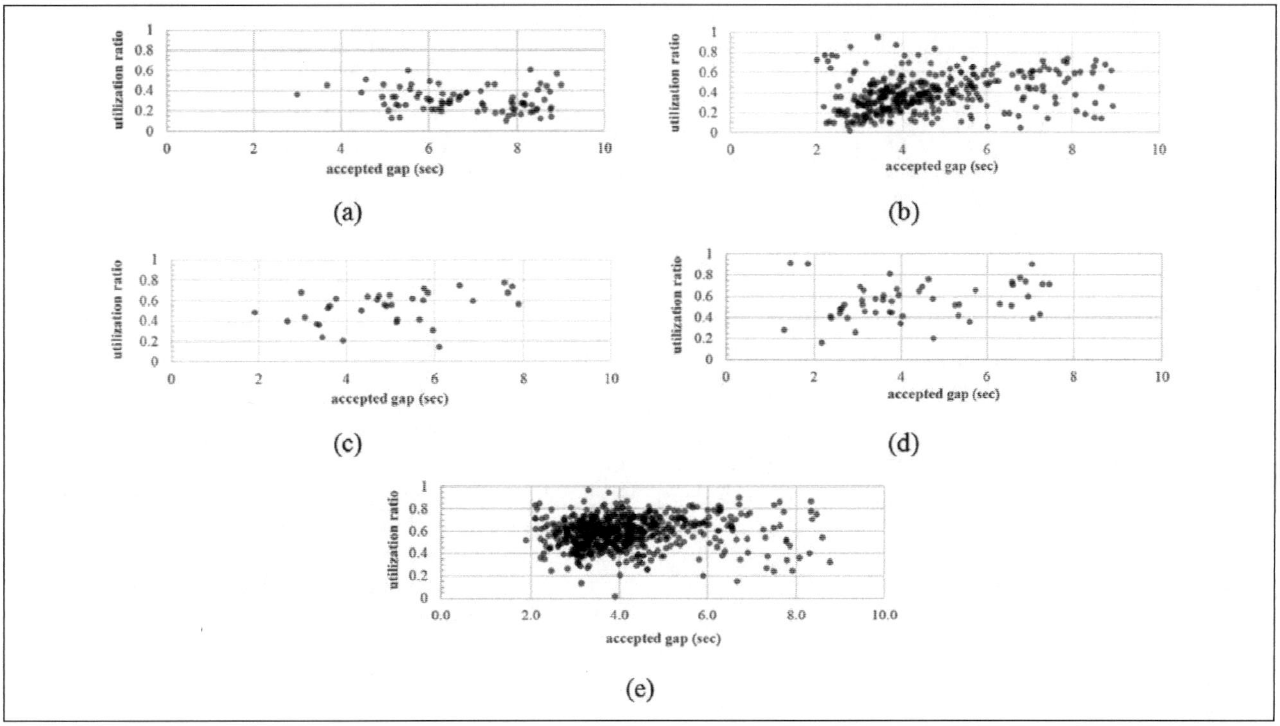

Figure 6. Accepted gap size plotted against gap utilization factor for the (*a*) Alaska, (*b*) Vermont, (*c*) New Hampshire, (*d*) Maine, and (e) New York roundabout locations.

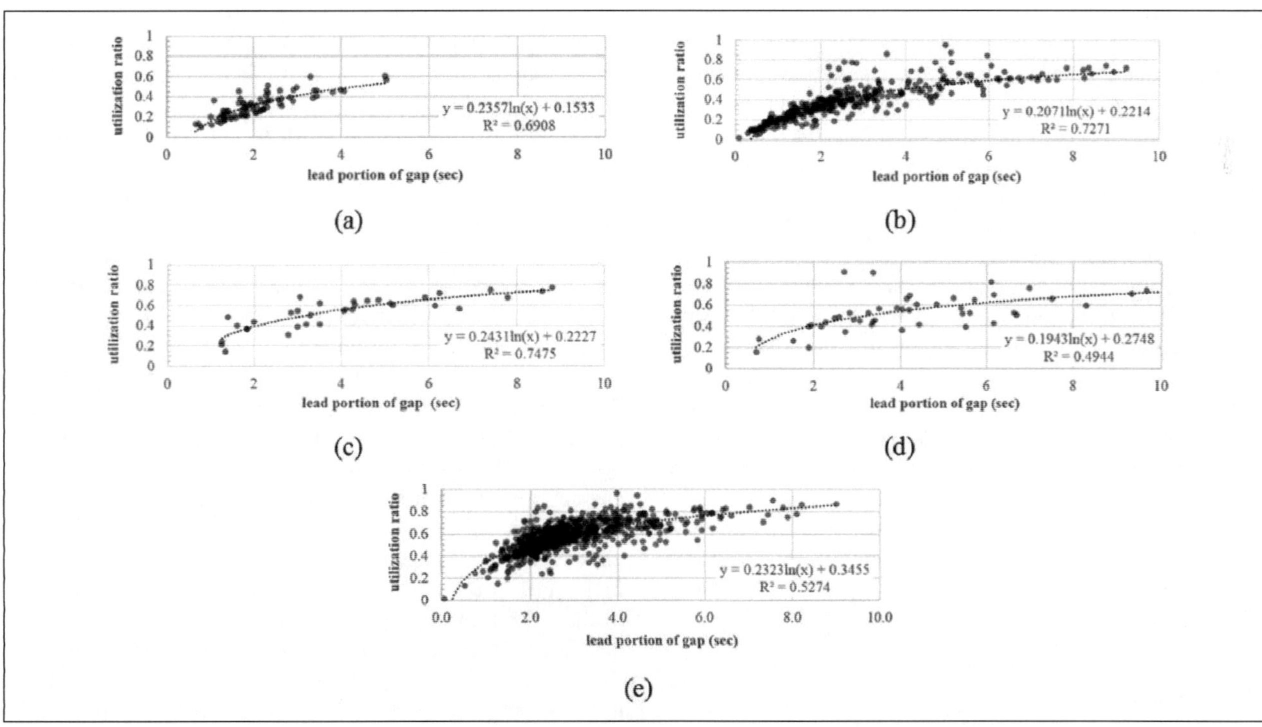

Figure 7. Lead portion of accepted gap plotted against gap utilization factor for the (*a*) Alaska, (*b*) Vermont, (*c*) New Hampshire, (*d*) Maine, and (e) New York roundabout locations.

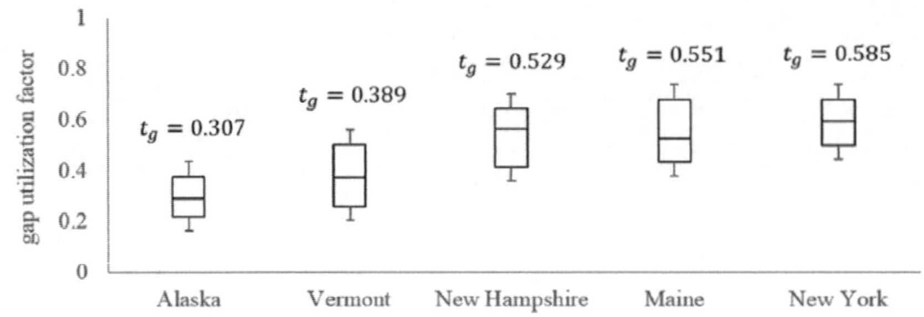

Figure 8. Between state comparison of gap utilization factor.

hypothesis that drivers would be more likely to use the lead portion of a gap (i.e., the usable gap would be skewed toward the leading vehicle in a circulating pair) as opposed to the lag portion. However, the logarithmic relationship coupled with the fact that the data points are more heavily-weighted low and left on the plots suggests that a larger percentage of accepted gaps are utilized closer to the lagging end of the gap (i.e., utilization values less than 0.5). Moreover, the gap utilization factor increases as the size of the accepted gap increases, which suggests that more of the lead portion of a gap is used when the length of the gap is relatively long.

An ANOVA test revealed that there are statistically significant differences in the average gap utilization across the five study states ($F = 131.6$, $d.f. = 4$, $p < 0.0001$). The between-state comparisons indicate that significant differences between the average gap utilization (Figure 8) existed for all pairs at $p < 0.001$, except between New York and Maine ($\mu_D = 0.034$, $p = 0.125$) and between Maine and New Hampshire ($\mu_D = 0.022$, $p = 0.507$). This may be partly attributed the fairly "flat" nature of the gap utilization distributions exhibited for Maine and New Hampshire as compared with the other three states.

A comparison of the gap utilization factors associated with priority-taking and non-priority taking events indicated that there is not a statistically significant difference between these groups ($t = 0.603$, $d.f.$, $= 97$, $p = 0.548$). The average gap utilization factor was also compared at roundabouts that had been in place for at least five years with those in place for less than five years. Previous studies have shown that at less than five years, there is a considerable amount of confusion regarding appropriate negotiation of roundabouts (*17, 24*). The results of the *t*-test indicate that roundabouts in place for five or more years have a statistically higher average gap utilization factor ($t = 9.59$, $d.f.$, $= 365$, $p < 0.0001$) with 0.544 and 0.419, respectively.

Discussion

The distribution of the data indicates significant temporal (time elapsed since construction) variations in the utilization of usable gaps by drivers, as well as spatial (between-site) differences across the study locations. New York has the highest average gap utilization factor, which fits with the expectation that drivers in this state tend to be more aggressive than drivers from other states. The comparatively lower average gap utilization factor for Alaska could imply a lack of familiarity with roundabout negotiation (there are fewer roundabouts in Alaska as compared with the other four states) or less aggressive tendencies (perhaps inherent to slower rates of acceleration from stop) of drivers. The latter is likely to be true for the Vermont roundabouts as well. This is partially supported by the lower limits of the plots for Alaska and Vermont as compared with the other three states as shown in Figure 6. Moreover, the nature of the logarithmic relationship indicates that a person who chooses to accept a relatively longer gap is more likely to enter on the leading portion of the gap, whereas a person who accepts a relatively smaller gap is more likely to enter into the lagging portion of the gap.

The original hypothesis was that a driver entering in the lagging portion of a gap would be more likely to influence the lagging vehicle of the circulating pair. This was expected to result in priority taking events (*17*). The lack of a statistically significant difference in this case could, in part, be because of the variability of drivers in the circulating vehicles, which was unaccounted for in this study. However, statistically significant differences exist between sites and between roundabout locations that have been in place for five or more years and those that have been in place for less than five years. This provides further evidence that as drivers become more familiar with roundabouts, the more efficient and appropriate their driving behavior becomes.

This paper presents a new gap-acceptance based parameter and novel method using real-world field data. The findings suggest potential merit in the consideration of this factor in the calibration and capacity analysis process for single-lane roundabouts. Follow-on work seeks to incorporate these findings into the HCM 2010 general roundabout capacity model as an additional parameter to the critical gap (t_c) and follow-up time (t_f) as a means of reducing overall error associated with between-site variations. This work builds on previous work and data collected by Belz et al. which identified non-compliant driver behavior at roundabout entries and by Belz which quantified within site variations in gap distributions and the effect of turn indicator usage on gap-acceptance (*17,18*).

One important limitation to note is that modern roundabouts are still relatively new to the United States. as compared with other countries. Very few sites operate at design year demands and as a result do not experience continuous saturated flow conditions, because of this, Lenters and Clayton note that a robust dataset of at-capacity observations is not achievable (*25*). Although the statistical analyses indicate significant spatial and temporal differences, more robust efforts should be made to ensure the results can be generalized while operating at sustained capacity, as well as incorporating more roundabouts across the United States to ensure that it will have broader geographic applicability.

References

1. Transportation Research Board. *Highway Capacity Manual 2010*. National Academy of Science, Washington, D.C., 2010.
2. Tanner, J. C. A Theoretical Analysis of Delays at Uncontrolled Intersections. *Biometrica*, Vol. 49, 1962, pp. 163–170.
3. Raff, M. S. *A Volume Warrant for Urban Stop Signs*. Saugatuck, Conn.: Eno Foundation for Highway Traffic Control, 1950.
4. Kimber, R. M. Gap-Acceptance and Empiricism in Capacity Prediction. *Transportation Science*, Vol. 23, No. 2, 1989, pp. 100–111.
5. Rodegerdts, L. A. *NCHRP Report 672 - Roundabouts: An Informational Guide*. Transportation Research Board of the National Academies, Washington, D.C., 2010.
6. Brilon, W., R. Koenig, and R. J. Troutbeck. Useful Estimation Procedures for Critical Gaps. *Transportation Research Part A: Policy and Practice*, Vol. 33, No. 3–4, 1999, pp. 161–186.
7. Cody, D., C. Mowakowski, and B. Benedicte. Observation of Gap Acceptance During Intersection Approach. *Proc., 4th International Symposium on Human Factors in Driver Assessment, Training and Vehicle Design*, Transportation Research Board National Research Council, Stevenson, Washington, D.C., 2007, pp. 321–327.
8. Jie, C., Y. Xinmiao, D. Wei, and H. Xin. Driver's Critical Gap Calibration at Urban Roundabouts: A Case Study in China. *Tsinghua Science and Technology*, Vol. 13, No. 2, 2008, pp. 237–242.
9. Kay, N., S. Ahuja, T. N. Cheng, T. van Vuren, and M. MacDonald. *Estimation and Simulation of Gap Acceptance Behaviour at Congested Roundabouts*. Association for European Transport, Strasbourg, France, 2006, pp. 1–16.
10. Tian, Z., R. Troutbeck, M. Kyte, W. Brilon, M. Vandehey, W. Kittleson, and B. Robinson. *Transportation Research Circular E-C018: Fourth International Symposium on Highway Capacity: A Further Investigation on Critical Gap and Follow-Up Time*. TRB, National Research Council, Washington, D.C., 2000. http://onlinepubs.trb.org/onlinepubs/circulars/ec018/ec018toc.pdf
11. Hagring, O. Estimation of Critical Gaps in Two Major Streams. *Transportation Research B*, Vol. 34, 2000, pp. 293–313.
12. Troutbeck, R. J. Traffic Interactions at Roundabouts. *Proc., 15th Australian Road Research Board (ARRB) Conference*. Australian Road Research Board, Victoria, Australia, 1990, pp. 17–42.
13. Alexander, L., P.-M. Cheng, M. Donath, A. Gorjestani, A. Menon, B. Newstrom, C. Shankwitz, N. Ward, and R. Starr. *Lag Acceptance Analysis for a Rural Unsignalized Intersection*. Presented at 86th Annual Meeting of the Transportation Research Board, Washington, D.C., 2007.
14. Macioszek, E. The Movement Model for Small Roundabouts with Minor Roads Capacity Estimating. *Transport Problems*, Vol. 1, 2007, pp. 87–94.
15. Polus, A., S. Lazar, and M. Livneh. Critical Gap as a Function of Waiting Time in Determining Roundabout Capacity. *Journal of Transportation Engineering*, Vol. 129, No. 5, 2003, pp. 504–509.
16. Kimber, R. *Traffic Capacity of Roundabouts, Laboratory Report 942*. Transport and Road Research Laboratory, Crawthorne, Berkshire, England, 1980.
17. Belz, N. P., L. Aultman-Hall, B. H. Y. Lee, and P. E. Gårder. An Event-Based Framework for Non-Compliant Driver Behavior at Single-Lane Roundabouts. *Transportation Research Record: Journal of the Transportation Research Board*, 2014. 2402: 38–46.
18. Belz, N. P. *Investigating the Role of Turn Indicator Usage by Exiting Vehicles in Gap Acceptance at Single-Lane Roundabouts*, Frontiers of Structural and Civil Engineering, Special Issue: Sustainable Civil Infrastructure, 2016.
19. Velan, S., and M. Van Aerde. The Impact of Driver and Flow Variability on Capacity Estimates of Permissive Movements. *Transportation Research Part A*, Vol. 32, No. 7, 1998, pp. 509–527.
20. Baek, S., P. Minnhagen, S. Bernhardsson, K. Choi, and B. J. Kim. Flow Improvement Caused by Agents Who Ignore Traffic Rules. *Physical Review E*, Vol. 80, 2009, pp. 016111.
21. Akcelik, R. *Transportation Research Circular E-C083: Roundabout Model Calibration Issues and a Case Study*. TRB, National Research Council, Washington, D.C., 2005. http://onlinepubs.trb.org/onlinepubs/circulars/ec083.pdf
22. Troutbeck, R., and S. Kako. Limited Priority Merge at Unsignalized Intersections. *Transportation Research Part A*, Vol. 33, No. 4, 1999, pp. 291–204.
23. Belz, N. P., L. Aultman-Hall, and J. Montague. Influence of Priority Taking and Abstaining at Single-Lane Roundabouts using Cellular Automata. *Transportation Research Part C: Emerging Technologies*, Vol. 69, 2016, pp. 134–149.
24. Retting, R., S. Krychenko, and A. McCartt. Long-Term Trends in Public Opinion Following Construction of Roundabouts. *Transportation Research Record: Journal of the Transportation Research Board,* 2007. 2019: 219–224.
25. Lenters, M., and R. Clayton. HCM Roundabout Capacity Methods and Alternative Capacity Models. *Institute of Transportation Engineers. ITE Journal*, Vol. 80, No. 7, 2010, pp. 22–27.

The Standing Committee on Roundabouts (ANB75) peer-reviewed this paper (18-04820).

Article

TRR
JOURNAL OF THE TRANSPORTATION RESEARCH BOARD

Transportation Research Record
2018, Vol. 2672(34) 10–19

DOI: 10.1177/0361198118758309
journals.sagepub.com/home/trr
SAGE

Impact of Geometric Factors on the Capacity of Single-Lane Roundabouts

Mark T. Johnson[1] and Ting-Li Lin[2]

Abstract

Roundabout capacity is primarily estimated by gap-acceptance or by geometric models. The 2010 Highway Capacity Manual (HCM 10) uses a gap-acceptance model developed by Siegloch with empirically derived values of critical gap and headway for single-lane and two-lane entries. A geometric capacity model was developed by Kimber and Hollis that diverged from gap-based models. The capacity of up to four lane entries was empirically derived from six geometric parameters. In 2012 capacity data were collected as part of the FHWA project, Assessment of Roundabout Capacity Models for the Highway Capacity Manual. These data were used to produce HCM6, an update of the HCM 2010. In HCM6, a capacity curve was fitted through all the capacity data for single-lane roundabouts. The large scatter of data about the mean capacity line suggests that the single-lane roundabouts may be separated into different geometric types to improve accuracy. To investigate this hypothesis, the capacity data were separated into two geometrically distinct types of single-lane roundabouts: smaller, compact and larger, curvilinear. The range of data for the disaggregated and aggregated data was compared. Also, a capacity line was derived for each type using the geometric capacity model and compared with the disaggregated data with the HCM6 method. The results demonstrate that differences in geometry, absent in HCM6, explain the wide data range. This was further confirmed by the geometric model that gave a good fit to both sets of data. These results indicate that the accuracy of capacity prediction is improved by including geometric variation.

US single-lane capacity data were collected in 2012 as part of the FHWA project to update HCM 10 (*1*) to the HCM6 (*2*). The capacity data were collected at 11 sites: five in Carmel, Indiana; two in New York State; three in Washington State; and one in Colorado.

Each data point is the observed saturated entry flow plotted against the observed circulating flow for periods of 1 min (Figure 1) (*3*). The data show a wide range in the observed capacities for all circulating flows. At circulating flows of 200 and 700 passenger cars per hour (pc/h), the observed entry capacities vary between 300 and 1,200 pc/h, and between 300 and 1,600 pc/h, respectively.

It is evident that an average capacity curve fitted to this wide range of data will not give good estimates of capacity for a given circulating flow. Consequently, there is a significant risk of overestimating or underestimating capacity, which may distort the design and evaluation process.

This wide range of capacity suggests that single-lane roundabouts may be separated into different geometric types.

To investigate this hypothesis, the observed capacity data were disaggregated into two distinct types of single-lane roundabouts—the smaller, more compact roundabouts and those with larger, more curvilinear geometry. Kimber's six geometric parameters allowed for distinguishing between the two types of roundabouts as described later in Table 3 and Figures 4 and 5.

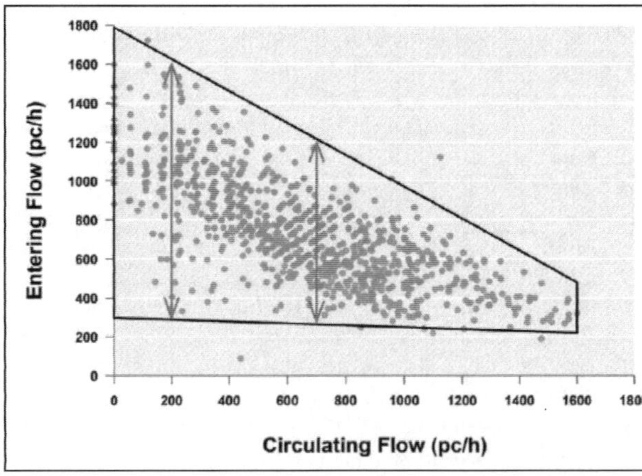

Figure 1. Aggregated single-lane field-collected HCM6 capacity data.

To test whether the derivation of capacity from the six geometric parameters gave a good fit to the data, the

[1]MTJ Roundabout Engineering, Madison, WI
[2]Briostat, Madison, WI

Corresponding Author:
Address correspondence to Mark T. Johnson: mark@mtjengineering.com

Table 1. Single-Lane Data Set in HCM6

Subarea	Site	Site code	Usable data points
Carmel, IN	136th Street/Keystone Parkway NB	IN12-E	4
Carmel, IN	136th Street/Keystone Parkway NB	IN12-S	71
Carmel, IN	96th Street/Ditch Road	IN07-E	31
Carmel, IN	96th Street/Ditch Road	IN07-S	10
Carmel, IN	96th Street/Westfield Boulevard	IN08-N	13
Carmel, IN	116th Street/Spring Mill Road	IN09-S	54
Carmel, IN	116th Street/Spring Mill Road	IN09-W	139
Carmel, IN	106th Street/Spring Mill Road	IN10-E	51
Carmel, IN	106th Street/Spring Mill Road	IN10-N	32
Carmel, IN	106th Street/Spring Mill Road	IN10-W	32
Carmel, IN	Subtotal	n/a	437
New York State	US 9/Warren Street/Hudson Avenue/Glen Street	NY07-S	71
New York State	US 9/Warren Street/Hudson Avenue/Glen Street	NY07-W	92
New York State	US 9/Warren Street/Hudson Avenue/Glen Street	NY07-E	65
New York State	US 9/Warren Street/Hudson Avenue/Glen Street	NY07-NW	13
New York State	US 9/Warren Street/Hudson Avenue/Glen Street	NY07-NE	1
New York State	Curry Road (SR 7)/Hamburg Street (SR14)	NY08-N	9
New York State	Curry Road (SR 7)/Hamburg Street (SR14)	NY08-W	21
New York State	Subtotal	n/a	272
Washington State	SR 166/Mile Hill Drive/Bethel Avenue	WA04-E	26
Washington State	SR 166/Mile Hill Drive/Bethel Avenue	WA04-N	16
Washington State	Lundeen Park Way/Callow Road	WA33-E	4
Washington State	Lundeen Park Way/Callow Road	WA33-W	13
Washington State	SR 20 Spur/Commercial Avenue	WA34-E	7
Washington State	SR 20 Spur/Commercial Avenue	WA34-N	5
Washington State	Subtotal	n/a	71
Colorado	Eby Creek Road/US 6 (Grand Avenue)	CO01-W	39
All Sites	Total data points	n/a	952
All Sites	Data points with one or more pedestrian events	n/a	133
All Sites	Total usable data points	n/a	819

uncalibrated geometric capacity model was used to predict the capacity line for each data set, and was then compared with the HCM6 exponential capacity lines fitted to each data set.

The 2012 single-lane data set shows that a significant quantity of data was collected from the five Carmel sites (437 usable data points), and also from the New York Glens Falls-US 9/Warren Street roundabout (242 usable data points) (Table 1) (*3*). This constitutes 72% of the total data collected for single-lane roundabouts.

The Glens Falls, NY-US 9/Warren St. roundabout and the Carmel roundabouts have distinctly different geometry. Glens Falls has smaller, more compact geometry, whereas the Carmel sites are large and curvilinear. The observed capacity data for each of these geometrically different types of roundabout were selected (Figure 2).

The data set for each site is distinct and the range of data for each geometric type is considerably less than the HCM6 aggregated data seen in Figure 1.

The relationship between the difference in the geometry and the difference in the capacity data between these two roundabout types will be examined in detail using both Kimber's uncalibrated geometric model and the uncalibrated and calibrated HCM6 model. However, this is preceded by a summary overview of each model.

The HCM6 calibrated model utilized the locally measured follow-on times from the New York and Carmel roundabouts (Table 2).

Overview of the Two Capacity Models

UK Geometric Capacity Model

The UK Transport Research Laboratory observed saturated capacity data at 86 roundabouts, collecting a total 11,000 min of data. They also conducted test track experiments on 35 geometric parameters to identify those that most affected capacity. Six of the 35 variables were found to significantly

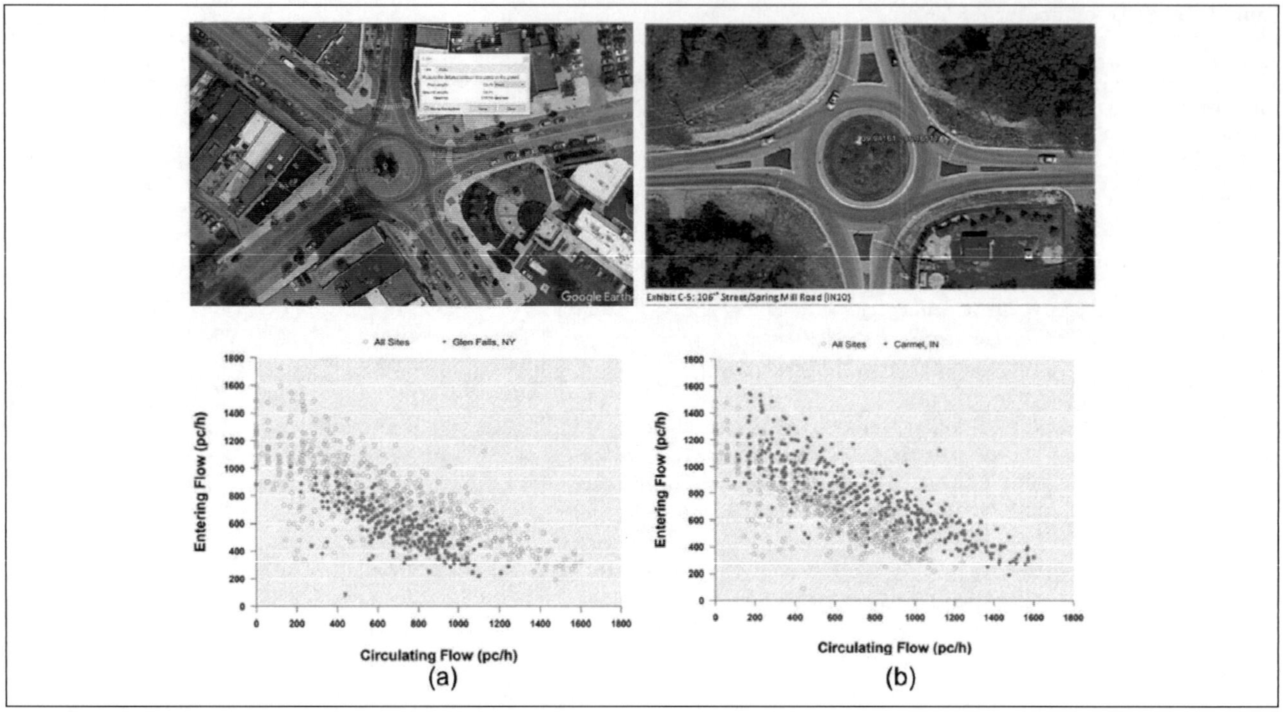

Figure 2. Capacity points for (*a*) smaller single-lane roundabout, Glens Falls, NY-US 9/Warren Street, and (*b*) representative larger curvilinear roundabout, Carmel, IN.

Table 2. Single-Lane Follow-on Times in HCM6

Site	Number of observations, n	Mean	Standard deviation
CO01-W	83	2.8	1
CO49-W2	4	2.9	0.9
IN07-E	40	2.5	1.1
IN07-S	38	2	0.8
IN08-N[a]	69	1.7	0.7
IN09-E	433	2.9	1.3
IN09-S	136	2.2	1.1
IN09-W	192	2.3	0.7
IN10-E	189	2.1	0.8
IN10-N	26	2.2	0.9
IN10-W	49	2	0.6
IN12-E	45	1.9	0.9
NY07-E	243	2.9	1
NY07-NW	141	2.8	0.9
NY07-NE	115	3	1.1
NY07-S	271	2.9	1
NY07-W	327	2.7	1.1

Averaged "local" follow-on times for Carmel and NY from HCM6 data:
IN07-E, S: 96th at Ditch Rd.: Ave. f = 2.26 s
IN08-N: 96th at Westfield: Ave. f = 1.70 s
IN09-S, W: Spring Mill Rd. at 116th: Ave. f = 2.2 s
IN10-E, N, W: Spring Mill Rd. at 106th: Ave. f = 2.09 s
IN12-E: Keystone Pkwy.: Ave. f = 1.90 s
NY07: Glens Falls/Warren/Hudson: Ave. f = 2.84 s

[a]IN09-E was removed from the weighted averages as there are no capacity data. Therefore, the follow-up time at IN09-E was not used for local calibration for Carmel sites. Similarly, this was applied to Kimber: the geometric parameters of IN09-E were not included.

Table 3. The Six Statistically Significant Variables of Roundabout Design

Geometric parameter	Symbol	Data range	Unit
Entry width[2]	e	3.6–16.5	Meters
Approach half width	v	1.9–12.5	Meters
Effective flare length	l'	1.0–∞	Meters
Entry radius	r	3.4–∞	Meters
Entry angle	ϕ	0.0–77	Degrees
Inscribed circle diameter	D	13.5–171.6	Meters

affect capacity of each entry (24 for a four-leg roundabout) (Table 3) (*4*).

The data from the 86 sites were used to directly relate capacity to each of the six geometric parameters. The result was Kimber's capacity model, named after the head of the research team. The model is linear, as it was not possible to demonstrate any degree of nonlinearity. The straight line fits accounted for more than 90% of the variance in the capacity.

Kimber's analytical framework was based on statistical principles, utilizing regression analysis to determine the most important geometric factors that have significant influence on the entry capacity. The model was derived as explicit functions of geometric factors in terms of the associations among them. The model coefficients were then jointly optimized by regressing the entry capacity on the independent variables.

The observed circulating flows Q_c were between 0 and 4,700 pc/h.

The entry capacity $Q_e = k(F - f_c Q_c)$
where

$$k = 1 - 0.00347(\phi - 30) - 0.978\left(\left(\frac{1}{r}\right) - 0.05\right),$$

$$F = 303x_2,$$

$$f_c = 0.210t_D(1 + 0.2x_2),$$

$$t_D = 1 + \frac{0.5}{1 + \exp\left(\frac{D - 60}{10}\right)},$$

$$x_2 = v + \frac{e - v}{1 + 2S},$$

$$S = \frac{1.6(e - v)}{l'},$$

and *e, v, l', D*, and *r* are in meters (or feet), ϕ in degrees, and Q_e and Q_c in pc/h.

These important geometric relationships have varying degrees of influence on capacity (Table 4). However, it is noted that the purpose of this study is not to examine the sensitivity of capacity to each geometric variable, but to determine if the geometric model may explain the wide variation in single-lane capacity data and lead to an improvement in capacity estimation (*5*).

HCM Model Development

The HCM 2010 roundabout capacity model was derived from directly-measured capacity data collected in 2003. A nonlinear (negative exponential) regression line was fitted to the data. The HCM 2010 recommends local calibration of capacity models to reflect local behavior as a means of providing more accurate capacity prediction. To accomplish this, it specifies a calibration procedure using the observed/measured average critical headway and follow-up headways. The *Y* intercept and exponential coefficient, or slope and curve of capacity line, are adjusted using these gap parameters.

$$Q_e = 1130.\exp(-0.0010Q\_c)$$

With local calibration, the equation becomes

$$Q\_e = A.\exp(-B.Q\_c)$$

where

$$A = \frac{3600}{t_f}$$

$$B = (t_c - t_f / 2) / 3600$$

t_f is the follow-up headway and t_c is the critical headway.

HCM 2010 Update: HCM6

The HCM 2010 update, HCM6, retains the 2010 model form (Siegloch) (*6*). The HCM6 capacity line was developed by fitting an exponential line to the data and then anchoring the *Y* intercept to a value derived from the average observed follow-up times of all the collected data. The average follow-up time was 2.6 s, and resulted in a *Y* intercept of 1,380 pc/h.

$$Qe = 1380.\exp(-0.00102Qc)$$

With local calibration, the equation becomes

Table 4. Geometric Model Variables

Three Major Capacity Variables

- Entry Width (E)
- Approach Width (V)
- Effective Flare Length (L')

Three Minor Capacity Variables

- Entry Radius (R)
- Entry Angle (ϕ)
- Diameter (D)

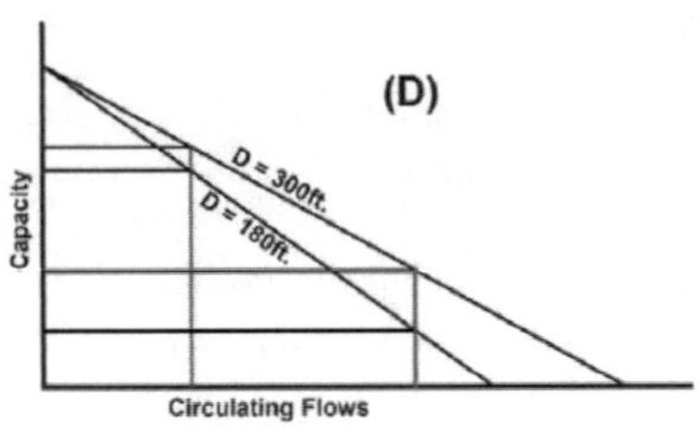

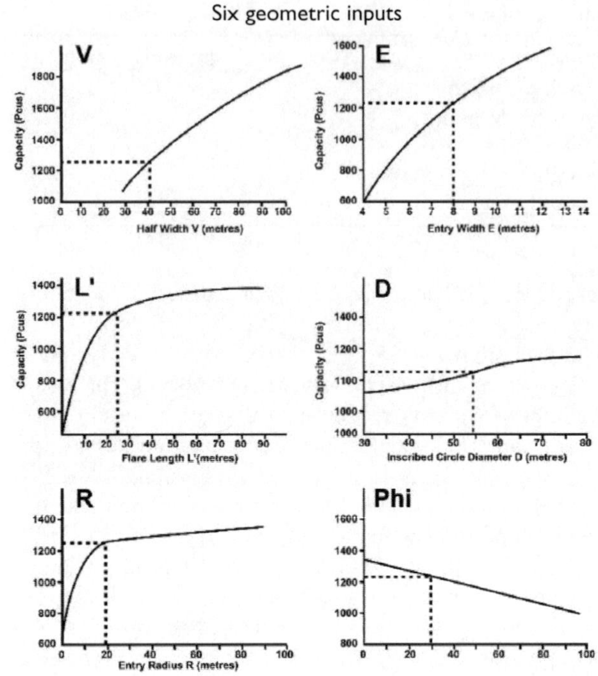

$$Qe = A.\exp(-0.00102Qc)$$

where

$$A = 3600 / t_f$$

t_f is the measured average follow-up time for that region.

Although the HCM 2010 used both the critical gap and the follow-up time for calibration to local conditions, the HCM6 study revised this, and concluded that there was weak correlation with the critical gap parameter. Calibration is now based only on follow-up time, which is now used to adjust the Y intercept of the capacity line. The unchanged capacity line is therefore moved either up or down with the adjusted Y intercept (*3*).

Assessment of the Form of the HCM Model

The HCM6 model does not relate driver behavior or capacity to roundabout geometry as their findings showed a low correlation between geometry and capacity.

The regression models for the single-lane roundabout sites show the fit to all of the single-lane data with exponential and linear models (Figure 3). Both of them fit equally well based on the root mean squared errors (RMSE). Figure 3 also shows the HCM6 model, which was calibrated to the average follow-up time, that is, the global Y intercept anchored at 1,380 pc/h. With this constraint the HCM6 model did not fit the data as well as the other two models. Statistically speaking, it was found that such a constraint was problematic and not well justified.

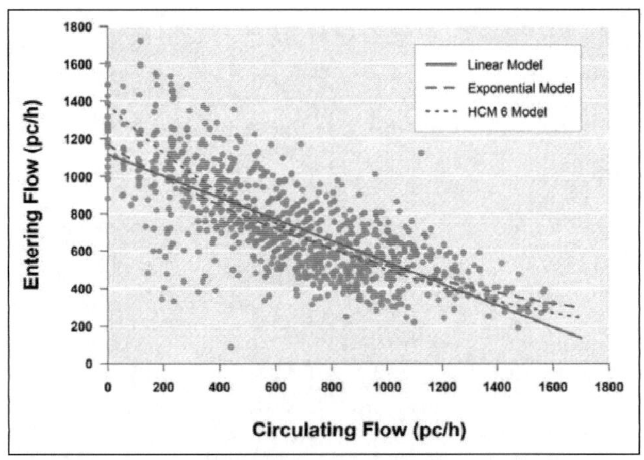

Figure 3. Regression models for single-lane roundabout sites: linear and exponential models fitted directly to the data; HCM6 model calibrated to average follow-up time.

Using regression analysis, fixing the Y intercept implies that it is a known value and there is no variation when X=0. This implies that there is no variation in capacity when the circulating flow is zero. However, this is not supported by the observed capacity variation at zero circulating flow. By

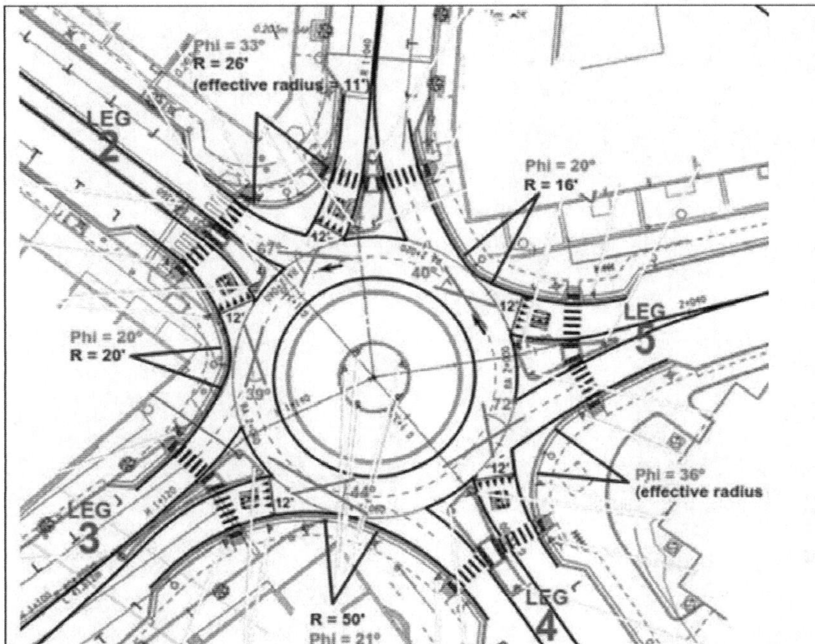

V	n		E	n	L'	R	Φ	D
11.00	1		12.00	1	20.00	21.00	26.00	105.00

Capacity Output

Y Intercept (vehs/hr)		Slope (tangent)		X Intercept (vehs/hr)	
Entry	Bypass	Entry	Bypass	Entry	Bypass
978		0.4845		2018	

Figure 4. Geometric parameters for Glens Falls, NY-US 9/Warren St./Hudson Ave./Glen St. roundabout.

anchoring the Y intercept, the HCM6 model ignored these variations and was unable to explain them.

Furthermore, a hypothesis test was performed to determine whether the Y intercept equals 1,385 based on the other two models. The calculated intercept of the linear model was found to be 1,117 with a standard error of 13.52, and the estimated intercept of the exponential model is 1,170 with a standard error of 1.02. Both models showed that the "true" intercept is significantly different from 1,385 (both p-values < 0.00001). Therefore, anchoring the Y intercept at 1,385 pc/h was not supported by the data or by statistical reasoning. In summary, both the linear and the exponential curves fit equally well to the aggregated single-lane data, and both performed better than the HCM6 anchored model that uses a global average follow-on time of 2.6 s.

Analysis of Data Segregated by Geometry

To assess if the geometric model may explain the wide variation in the data, the HCM6 field data were disaggregated into two data sets based on two distinct types of single-lane geometry: the smaller, compact geometry at the Glens Falls, NY, site and the larger, more curvilinear geometry at the five Carmel roundabouts (Figures 4 and 5). The disaggregated data formed two distinct groups with less spread when compared with aggregated HCM6 data.

Kimber's six geometric capacity variables were measured for the two types of roundabouts. As each type has a range of geometry, the six geometric parameters were averaged for each type.

(a)

(b)

(c)

(d)

(e)

	Carmel	
	Range	Avg
e'	13-15	14
v	0	12
l'	20-25'	23'
D	130-148'	138
r	55-70	65
Phi ϕ	12-20	16 deg

Geometric Inputs (Averaged for All Entries)

V	n	E	n	L'	R	Φ	D
12.00	1	14.00	1	23.00	65.00	16.00	138.00

Capacity Line Output

Y Intercept (vehs/hr)		Slope (tangent)		X Intercept (vehs/hr)	
Entry	Bypass	Entry	Bypass	Entry	Bypass
1298		0.5783		2244	

Figure 5. Carmel, IN, roundabouts and geometric inputs: (*a*) 116th/Spring Mill Rd., (*b*) 106th/Spring Mill Rd., (*c*) 96th St./Ditch Rd., (*d*) 96th St./Westfield Blvd., and (e) 136th St./Keystone Pkwy.

Smaller Single-Lane Geometrics: NY

The geometric parameters for the NY roundabout and its averaged values are illustrated in Figure 4. These are: approach roadway width (v) = 11', entry width (e) = 12', flare length (l') =20', entry radius (r) = 21', entry angle (ϕ) = 26°, and diameter (D) = 105'.

Larger, Curvilinear Single-Lane Geometrics: Carmel

The five larger, more curvilinear roundabouts in Carmel, IN, are: 116th St. & Spring Mill Rd., 106th St. & Spring Mill Rd., 96th St. & Ditch Rd., 96th St. & Westfield Blvd., and 136th St. & Keystone Pkwy. (Figure 5).

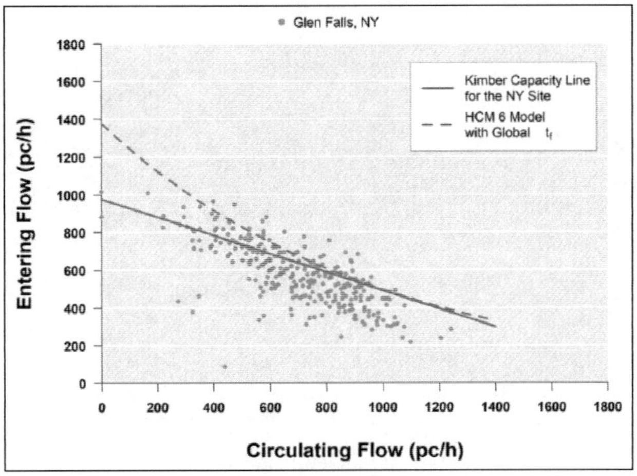

Figure 6. Glens Falls, NY: Uncalibrated Kimber capacity line ($Y = 995 - 0.4695X$) for the NY site compared with uncalibrated HCM6 model with global follow-on time of 2.6 s.

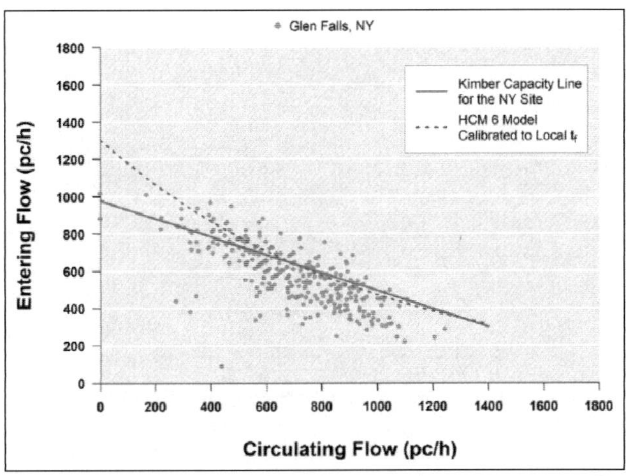

Figure 7. Uncalibrated Kimber capacity line ($Y = 995 - 0.4695X$) for the NY site compared with HCM6 calibrated with local follow-on of 2.8 s.

Figure 5 includes the geometric parameters for each roundabout and their averaged values. These are: approach road width (v) = 12', entry width (e) = 14', flare length (l') = 23', entry radius (r) = 65', entry angle (ϕ) = 16°, and diameter (D) = 138'.

Using the average geometric parameters of each type, the predicted capacity line of Kimber's uncalibrated model was compared with the field-measured data and this was compared with the anchored HCM6 capacity line, calibrated to the local follow-on time.

The RMSE value was derived for each model to assess how closely the predicted capacity line fits the observed data. The model with the lower RMSE is the better fit. (The statistical analysis was performed using R Version 3.2.3). The results of this analysis are discussed in the following sections.

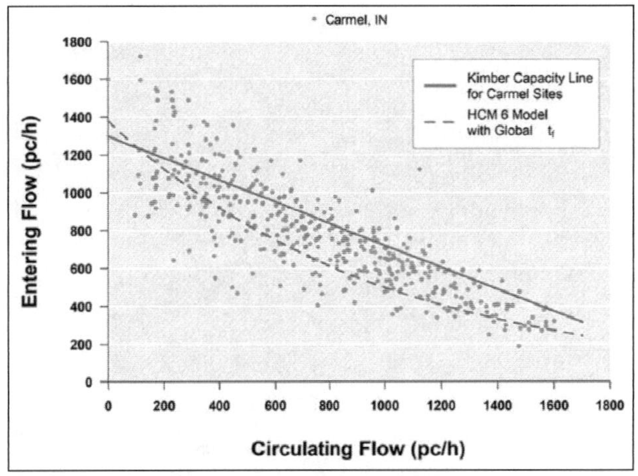

Figure 8. Uncalibrated Kimber capacity line ($Y = 1298 - 0.5783X$) for the Carmel sites compared with HCM6 model uncalibrated with global follow-on time of 2.6 s.

Analysis Results of the Smaller, Compact Geometry: NY Glens Falls Data

Kimber's uncalibrated model used the average geometric parameters to predict the capacity. The capacity line was also predicted by the uncalibrated HCM6 model using the global follow-up time (2.6 s). Both lines were compared with the observed capacity data (Figure 6).

The RMSE values for uncalibrated Kimber and uncalibrated HCM were 112 and 164, respectively. Kimber's geometric model was a better match to the data than HCM6 which overestimated the capacity at low circulating flows. The geometric model's prediction suffers from using averaged geometric parameters. When the actual parameters are used for an entry, the fit to the observed capacity data for that entry is much better.

Kimber's uncalibrated geometric model is compared with the calibrated HCM6 model (2.8 s local follow-on time) with the observed capacity data (Figure 7).

The RMSE for the Kimber uncalibrated model was unchanged at 112, and the RMSE for HCM6 calibrated to the average local follow-up time (2.8 s) was 138. The calibrated HCM6 model had a better fit than the uncalibrated HCM6, but it was not as good as Kimber's uncalibrated model. It also continued to overestimate the capacity at low circulating flow.

Analysis Results of Larger, Curvilinear Geometrics: Carmel Data

Kimber's uncalibrated geometric model is compared with the uncalibrated HCM6 model (2.6 s global follow-on time) with the observed capacity data (Figure 8).

The RMSE for the uncalibrated Kimber model was 172. The uncalibrated and locally calibrated HCM6 models had an RMSE of 183 and 178, respectively. The geometric model

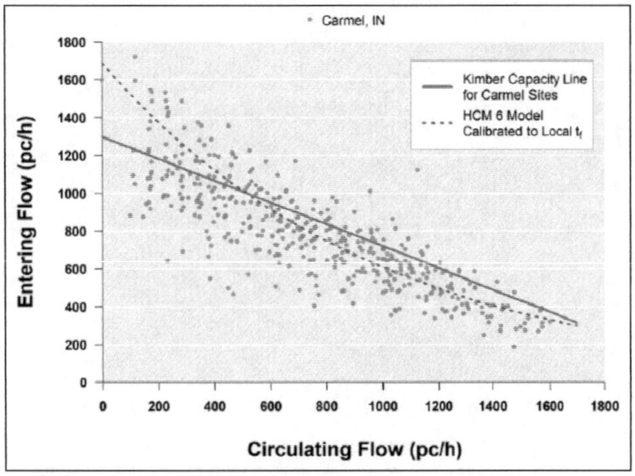

Figure 9. Uncalibrated Kimber capacity line ($Y = 1298 - 0.5783X$) for the Carmel sites compared with HCM6 model calibrated with local follow-on time of 2.13 s.

Table 5. Averaged Data and Geometric Inputs for 106th and 116th at Spring Mill Rd.

Parameter	Range	Average
e	0	14
v	0	12
l'	20–30'	25'
D	140–148'	144
r	55–67	62
ϕ	15–21	19°

performed slightly better than both of the HCM6 models on this segregated data set.

Kimber's uncalibrated geometric model is compared with the locally calibrated HCM6 model (2.13 s local follow-on time) with the observed capacity data (Figure 9).

The RMSE for the locally calibrated HCM6 model was 178. As noted previously, the geometric model performed slightly better than both of the HCM6 models on this segregated data set.

Further Data Reduction by Geometric Variables

Although the geometric parameters among the five Carmel sites were similar, they still presented a range of geometries. Subsequent evaluation of the data reveals that of the five Carmel sites, 116th at Spring Mill Rd. and 106th at Spring Mill Rd. have less geometric variation though they still include a significant proportion of the Carmel data.

To determine if the geometric model provided a better fit with a narrower geometric range, the capacity data from these two sites were examined. The average geometric parameters for the two Carmel roundabouts are compared with the average of all five sites—the average radius (r) decreased, phi angle (ϕ) increased, and diameter (D) increased (Table 5).

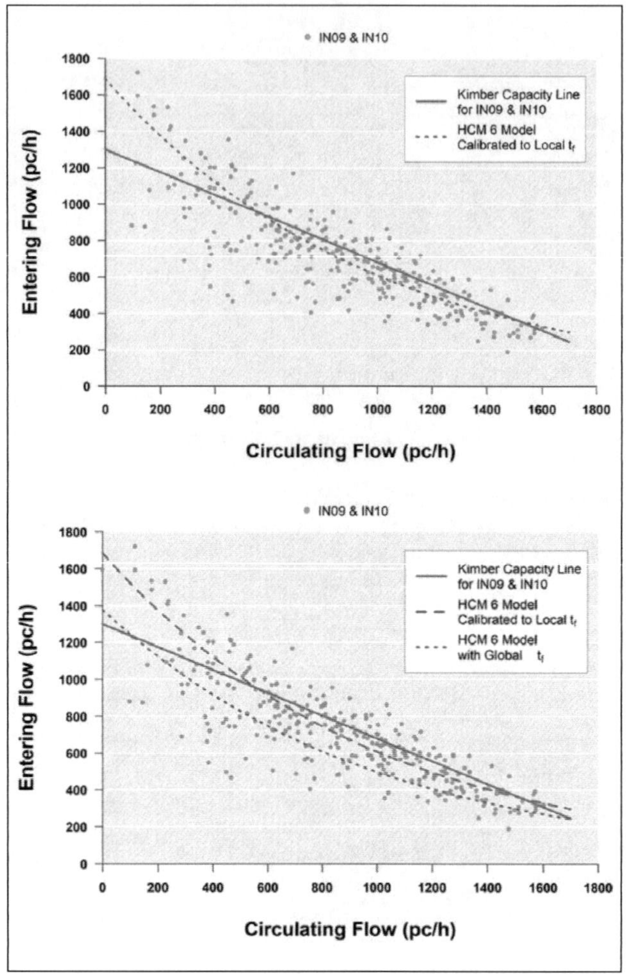

Figure 10. Uncalibrated Kimber capacity line ($Y = 1300 - 0.6198X$) for the two Carmel sites compared with uncalibrated and calibrated HCM6 models (follow-on times of 2.6 s and 2.13 s, respectively).

The uncalibrated Kimber model (using the average geometry of the two sites) is compared with both the uncalibrated and locally calibrated HCM6 models with the observed capacity data from these two roundabouts (Figure 10).

The RMSE for Kimber's uncalibrated model was 155. The uncalibrated and locally calibrated HCM6 models had an RMSE of 181 and 158, respectively. Kimber's model performed better with the narrowed geometric parameters than both of the HCM6 models. Narrowing the data from the five Carmel sites to two sites reduced the uncalibrated Kimber RMSE from 172 to 155.

Summary

Kimber's uncalibrated geometric capacity model gave better predictions than the locally calibrated HCM6 model. Without local calibration the geometric model provided good predictions of capacity at the geometrically distinct Glens Falls, NY, site and the Carmel sites. This implies that the large variation in the data from all sites is caused by geometric variation. The results show that with smaller, more compact roundabouts, the HCM6 model overestimated the entry capacity; with larger, more curvilinear roundabouts, however, it underestimated the entry capacity at mid- to higher-range circulating conditions and overpredicted capacity at lower circulating flows.

The HCM6 model with local calibration performed better. However, it is noted that the analysis used local follow-on times for roundabouts with distinct geometries. This suggests that the differences in geometry are a significant cause of the different follow-on times for each location. This was reinforced by the results from the uncalibrated geometric model. It is therefore concluded that the difference in driver behavior, manifested by the difference in follow-on times at each site, is caused primarily by the differences in geometry.

Conclusion and Future Research

This paper examined the single-lane capacity data collected in 2012 for existing saturated single-lane US roundabouts as part of the FHWA project, Assessment of Roundabout Capacity Models for the Highway Capacity Manual. These data were used to update HCM 2010 to HCM6.

The wide range of single-lane data is problematic for accurate capacity prediction. This wide scatter suggests that there may be geometric distinctions between single-lane roundabouts that affect their capacity. To examine this hypothesis, the data were segregated into two geometrically distinct types of roundabouts—smaller, compact geometrics and larger, curvilinear geometrics. The data points for each type of single-lane roundabout were distinct and had less scatter than the aggregated data of all the single-lane roundabouts.

To further examine if these differences in capacity were caused by the differences in geometry, Kimber's uncalibrated geometric model was used to predict the capacity line using the geometry of each type. This gave a good fit to both sets of data, reinforcing the hypothesis that the capacity variation is explained by the difference in geometry.

The HCM6 model relies on local follow-on times to account for differences in driver behavior. However, Kimber's geometric model relates capacity directly to geometry. The good fit between the uncalibrated geometric model and the local data indicates that the geometric differences between sites are the primary cause of the difference in driver behavior between sites.

These findings could be further tested by selecting geometrically opposite roundabouts for each of the two regions (a larger, more curvilinear roundabout in New York, and a smaller, more compact single-lane roundabout in Indiana). If the results support those of this paper, it would confirm that the differences in driver behavior are caused primarily by differences in geometry rather than "regional" nongeometric driver behavior. Roundabouts in the same area with markedly different geometries would also be a useful test, as the regional follow-on time would be the same for both, but the actual times quite different because of their geometric differences.

Acknowledgments

The authors would like to thank the FHWA and the HCM6 research team for providing data utilized in this paper and for their willingness to answer questions and provide information.

References

1. *Highway Capacity Manual 2010.* Transportation Research Board of the National Academies, Washington, D.C., 2010.
2. *Highway Capacity Manual: A Guide for Multimodal Mobility Analysis*, 6th ed. Transportation Research Board, Washington, D.C., 2016.
3. *Accelerating Roundabout Implementation in the United States: Assessment of Roundabout Capacity Models for the Highway Capacity Manual.* FHWA-SA-15-070, Volume II. FHWA, U.S. Department of Transportation, 2015.
4. Kimber, R. M. *The Traffic Capacity of Roundabouts.* TRRL Laboratory Report 942. Transport and Road Research Laboratory, Crowthorne, Berkshire, 1980.
5. Kimber, R. M. Gap-Acceptance and Empiricism in Capacity Prediction. *Transportation Science*, Vol. 23, No. 2, 1980, pp. 100–111.
6. Siegloch, W. Die Leistungsermittlung an Knotenpunkten ohne Lichtsignalsteuerung (Capacity evaluation at unsignalized intersections). *Strassenbau und Strassenverkehrstechnik*, No. 154, 1973.

The Standing Committee on Roundabouts (ANB75) peer-reviewed this paper (18-01538).

Article

TRR
JOURNAL OF THE TRANSPORTATION RESEARCH BOARD

Transportation Research Record
2018, Vol. 2672(34) 20–28

DOI: 10.1177/0361198118794055
journals.sagepub.com/home/trr
SAGE

Roadway-Related Truck Crash Risk Analysis: Case Studies in Texas

Qun Zhao[1], Tyrie Goodman[1], Mehdi Azimi[1], and Yi Qi[1]

Abstract

Since 1994, Texas has had the highest number of fatal crashes involving large trucks in the United States, and this number increased by 82% from 2009 to 2012. Due to the size and weight of large trucks, their crashes usually are very destructive. Although large trucks have a significant impact on traffic safety in Texas, very little analysis has been conducted of the risk factors associated with crashes involving large trucks, especially the roadway-related risk factors. In this paper, the results of a collision-diagram-based analysis are presented for selected areas in Texas where frequent crashes of large trucks occur. First, historical data related to large truck crashes from 2011 through 2015 were extracted and entered into ArcGIS to identify areas within a 0.5-mi radius of where large truck crashes occur frequently, which were named hot spots. Then, based on the results of the identified hot spots, we identified hot areas, that is, areas with clusters of hot spots. Police reports of all of the crashes that occurred in the selected hot areas were then reviewed, and collision diagrams were developed. By analyzing all of the collision diagrams that were developed, five roadway-related risk factors were identified, and potential effective countermeasures were proposed to prevent or mitigate crashes involving large trucks.

A large truck, as defined by the U.S. Department of Transportation, is any vehicle with a gross weight greater than 10,000 lb. Since 1994, Texas has had the highest number of fatal crashes involving large trucks in the United States. Due to their size and weight, crashes involving large trucks often result in fatal injuries, property damage, and a significant amount of economic loss. According to the Texas Department of Transportation's (TxDOT) Texas Motor Vehicle Crash Statistics, from 2009 to 2012, the number of fatal crashes involving large trucks statewide increased from 299 to 543, an 82% increase. Crashes involving large trucks are a major health and safety concern for Texans.

Various factors can contribute to crashes involving large trucks, for example, roadway conditions, driver-related factors, vehicle-related factors, and environmental conditions. Among these factors, the roadway-related factors are more controllable by traffic engineers/roadway designers, and the frequency of such crashes can usually be reduced by providing better roadway geometric design or using better traffic control strategies. In addition, as large trucks take much more time and distance to accelerate, make turns, and stop than other vehicles, factors related to the road infrastructure, such as the roadway geometric conditions and traffic control methods, have more significant impacts on crashes involving large trucks than other types of crashes. Thus, the risk factors associated with the roadway must be investigated so that roadways can be designed that are safer and provide better accommodation of the movement of large trucks. In this study, the roadway-related factors that contribute to the crashes of large trucks were investigated by conducting a collision-diagram-based case study at selected hot areas of large truck crashes in Texas. By analyzing the collision diagrams that were developed, five roadway-related risk factors were identified and possible countermeasures were proposed. The results of these studies will provide traffic engineers/roadway designers with a better understanding of the roadway-related risk factors contributing to large truck crashes and assist them in making roadways less risky for large trucks.

Literature Review

Extensive previous research has been conducted to investigate the roadway-related factors that contribute to crashes involving large trucks. These research topics include the grade of the roadway, curves, type of road,

[1]Texas Southern University, Houston, TX

Corresponding Author:
Address correspondence to Yi Qi: qiy@tsu.edu

posted speed limit, lighting conditions, and the condition of the pavement.

Previous research has shown that roadway-related factors were present for 16% of two-vehicle crashes involving large trucks (*1*). Several geometric characteristics also were found to have relationships with the crash rate, such as the number of lanes, lanes less than 11-ft wide, the width of the shoulders, and grades greater than 4% (*2*).

According to a study conducted by Jerry and Agent, specific roadway areas can be identified as high-risk locations for the crashes of large trucks, that is, interchange ramps, steep grades, sharp curvatures on two-lane roads, intersections with restricted sight distance, and high speed (*3*).

Crashes involving large trucks occur on rural highways and in more congested roads in urban areas. In addition, the geometric design, type of road, and the traffic control measures implemented on a road can affect the results of crashes of large trucks, for example, limiting them to minor rear-end collisions or perhaps resulting in multi-vehicle crashes with fatalities. Road alignment is important because it can affect both the risk of crashes and their severity. Garber and Joshua indicated that 98% of fatal crashes due to drivers' errors occur on curves, which is an alarming figure that indicates that the relationship between roadway conditions and the occurrence of crashes involving large trucks should be investigated further (*4*).

Burke et al. considered the effects of roadway lighting, among other factors, on the severity of crashes (*5*). Crashes that occurred at night on a roadway with street lighting were predicted to be less severe. A driver who has increased night vision due to lighting may have a heightened ability to avoid a crash or minimize its severity.

Most of the previous research used statistical modeling or a variable-orientated approach to identify the factors that have significant impacts on the frequency and severity of the crashes involving large trucks. For these studies, to develop a statistical model, initially, a set of variables must be created based on the information contained in the crash database. Usually, crashes are recorded using a standardized format and only a limited set of attributes can be included. As a result, the statistical models that have been developed can only identify a limited set of risk factors, which are usually quite generic, such as speed limits, curves, and street lighting. Some specific contributing factors at the local level, such as a specific traffic control device at the crash site or a specific geometric condition, cannot be taken into account easily by statistical models. This type of variable-orientated approach has been known to have the limitation of being "not very in-depth" (*6*). To fill this gap, in this study, we conducted a risk analysis based on collision diagrams. Initially, hot areas with high truck-crash frequencies were identified based on historical crash data collected through the state of Texas. Then, a crash diagram was developed to further identify the spots where the same or similar types of crashes had occurred many times. Subsequently, police crash reports of the individual crashes were examined carefully to identify the specific crash risk factors at these spots. This approach combines the strength of both large crash data analysis and detailed case-study approaches, making the research findings more unique and comprehensive.

Data Collection and Process

The data used in this research were obtained from TxDOT's Crash Records Information System. Records of crashes involving large trucks during the 5 years from 2011 through 2015 were extracted and analyzed by using a geographic information system (GIS). Table 1 shows the number of crashes in each of the years and their severity levels.

The GIS approach to identify hot spots where crashes involving large trucks occurred includes the following key steps:

Step 1: Input Crash Data into GIS

The extracted data for crashes involving large trucks have longitude and latitude information for every crash

Table 1. Crashes Involving Large Trucks in Different Years and the Related Crash-Severity Levels

	Number of crashes of large trucks				
Year	Fatal	Incapacitating	Non-incapacitating	Possible injury	Total
2015	465	968	2,648	3,661	7,742
2014	466	955	2,494	3,347	7,262
2013	419	874	2,346	2,980	6,619
2012	425	764	2,200	2,950	6,339
2011	345	651	1,977	2,786	5,759
Total	2,120	4,212	11,665	15,724	33,721

record. Therefore, the data can be easily converted to a GIS layer. In addition, the layer of a Texas roadway network was also inputted into a GIS map.

Step 2: Apply the "Integrate" Tool to Integrate the Closely Located Crashes to One Hot Spot

"Integrate" is a tool of ArcGIS, and can be used to maintain the integrity of the shared feature boundaries by making features coincident if they fall within the specified *x,y* tolerance. Features that fall within the specified tolerance are considered identical or coincident. Integrate accepts only simple feature classes as inputs (point, multipoint, line, or polygon). In ArcGIS, the Integrate tool performs the following two tasks:

1) Identifies features that are within the given tolerance distance
2) Inserts common coordinate vertices for features that fall within the given *x,y* tolerance

In Figure 1, which shows an example of integrating crash data, there are five crashes, one each at points A, B, C, D, and E, and the distances between nearby points are marked. If a tolerance distance of 0.5 mi is assumed, by running the Integrate tool, points A and B would be integrated because they are within the tolerance distance of 0.5 mi. C, D, and E also would be integrated.

In this step, the value of the tolerance distance is critically important. A tolerance distance that is too large may collect all the accidents spread over a large area; as a result, the hot spots tend to be at the locations with high volumes of truck traffic, such as the areas around port terminals or truck distribution centers. A tolerance distance that is too small may split the whole area into many small areas with only a few points falling into each area. Due to the random nature of traffic accidents, a small sample size cannot be used to determine the true hot spots. In this study, we tried using different distances, and we found that a tolerance distance of 0.5 mi produced reasonable results for hot spots. In addition, in most transportation safety studies, roadways usefully are split into 1-mi segments for crash risk analysis. Using a tolerance distance of 0.5 mi will integrate the crashes within the 1-mi distance, which is consistent with traditional safety studies.

Step 3: Apply the "Collect Events" Technique to Count the Crashes at Each Hot Spot

The "Collect Events" tool combines the integrated points. It creates a new output feature class containing all of the unique features in the integrated feature class that was used as input. Then, it adds a field named "ICOUNT" to give the sum of all of the crashes at each hot spot. In this study, the top 20 spots were selected for each year.

Based on the results of the identified hot spots, the hot areas with clusters of hot spots were identified. After the hot areas were identified, we collected all of the police reports for those hot areas from 2011 through 2015 for further review.

Methodology

In this study, an analysis based on the collision diagrams was conducted to identify the roadway-related factors that contribute to the crashes of large trucks. To be more specific, collision diagrams were created for each study area to identify the patterns and locations where certain types of crashes occurred frequently.

To develop a collision diagram, the first step is to draw the background map. After that, detailed police reports of crashes that occurred in this area were examined carefully. The locations of the crashes were marked

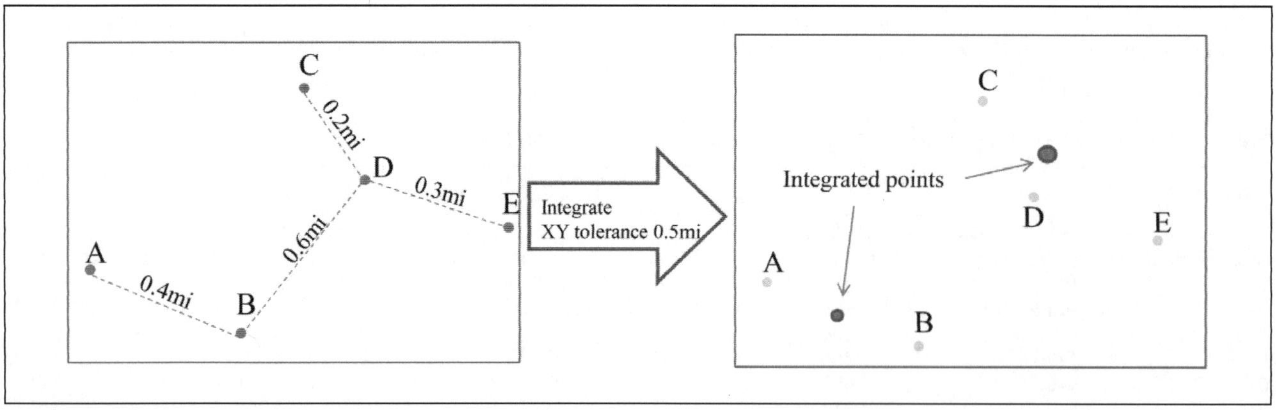

Figure 1. Example of integrating crash data.

Table 2. Symbols and Definitions of Types of Crashes in the Collision Diagrams

Type	Symbol	Definition
Rear-end		The front of one vehicle collides with the rear of another vehicle while both vehicles are traveling in the same direction
Sideswipe	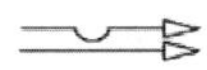	One vehicle impacts another traveling in the same direction by "swiping" along the surface with the direction of travel
Dual left-turn		Two vehicles collide with each other when both make left turns
T-bone		One vehicle disregards a signal or sign and collides with a crossing vehicle
Angle	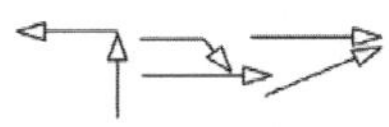	One through vehicle collides with another through vehicle in the crossing direction, or two vehicles collide while traveling in the same direction while changing lanes
Right-turn angle	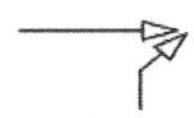	One right-turning vehicle collides with another vehicle on the cross street
Left-turn angle		An accident involving two opposing vehicles when one of them is turning left
Back		An accident involving two vehicles traveling in the same direction when the leading vehicle backs up
Object		One vehicle hits a roadside object (e.g., the curb)
Pedestrian		An accident in which a vehicle and a pedestrian collide

on the map with different symbols representing different types of crashes.

There were eight major types of crashes identified in the collision diagrams, that is, rear-end, sideswipe, angle, right-turn angle, left-turn angle, back, object, and pedestrian. The definitions of these types of crashes along with their symbols are given in Table 2.

The samples of the collision diagrams that were developed are presented in the results section below.

Analysis of Results

After analyzing each collision diagram, five major roadway-related risk factors were identified:

1. Intersection with dual left-turn
2. Insufficient turning space for trucks making right turns
3. Two-way frontage road design
4. Short merge distance at highway entrances
5. Failure to control speed on highway exit ramps

In the following section, each risk factor is explained, and the sample crash report and collision diagram are provided.

Intersection with Dual Left-Turn and Heavy Truck Volume

A left-turn is provided at intersections to improve the safety of the intersection and vehicle operations by separating left-turning vehicles from through vehicles. At intersections with extensive turning movements, it may be necessary to consider providing multiple turn lanes to accommodate the increased demand for turning. Dual left-turn lanes are commonly used for this purpose (*7*).

Dual left-turn lanes do not pose a threat to safety during normal traffic conditions. However, this design may

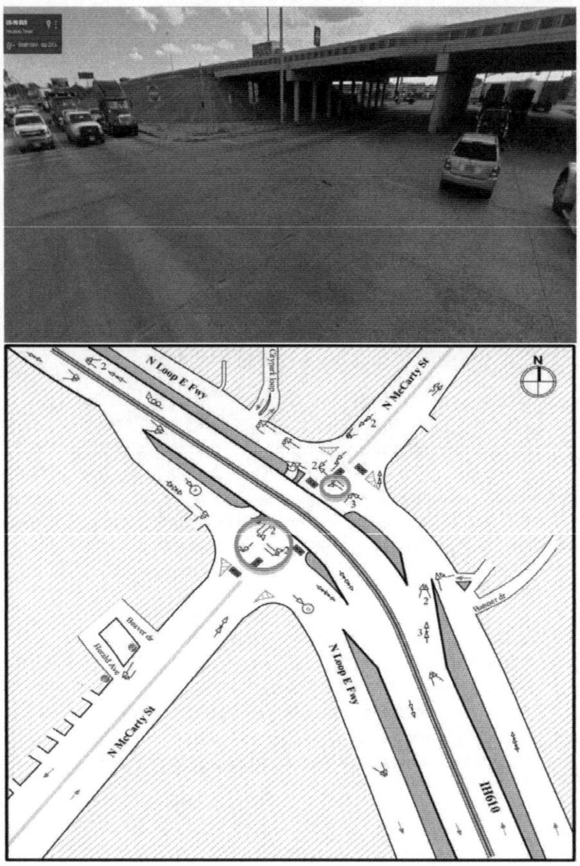

Figure 2. Collision diagram for large truck crash hot area "Houston IH610 @ US90."

be problematic when there is a heavy volume of truck traffic. Similar crashes occurred in the identified hot areas in Houston, Laredo, and South Dallas. Figure 2 shows a collision diagram developed for a hot area in Houston, and the Google Maps street view for this location.

Five double left-turn crashes occurred at this intersection, as circled in Figure 2. After examining this intersection, we found that a dual left-turn setting applies for all four directions. Heavy truck volume makes the dual left-turn movement riskier. In addition, the turning guide lines on the pavement were almost worn off the road, and the signage to assist drivers also was lacking.

To reduce double left-turn crashes, pavement markings should be applied for turn lanes to clearly designate the path to be taken by vehicles. In addition, designated signal light heads for each lane (as shown in Figure 3) can also be implemented at this type of intersection to assist drivers making left turns.

Insufficient Turning Space for Trucks Making a Right Turn

There are two issues for trucks making a right turn. First, the lengths of the trucks require the drivers to make wide turns, causing them to encroach upon adjacent lanes of traffic. In addition, trucks take a longer time to complete turning, thus they have longer exposure times to hazards during a right turn on red.

It was noticed that insufficient turning spaces appear in many places in Texas, such as in Austin, Laredo, and

Figure 3. Sample of designated signal light heads for each lane.

Figure 4. Sample crash and Google Map of one hot area (Austin, TX).

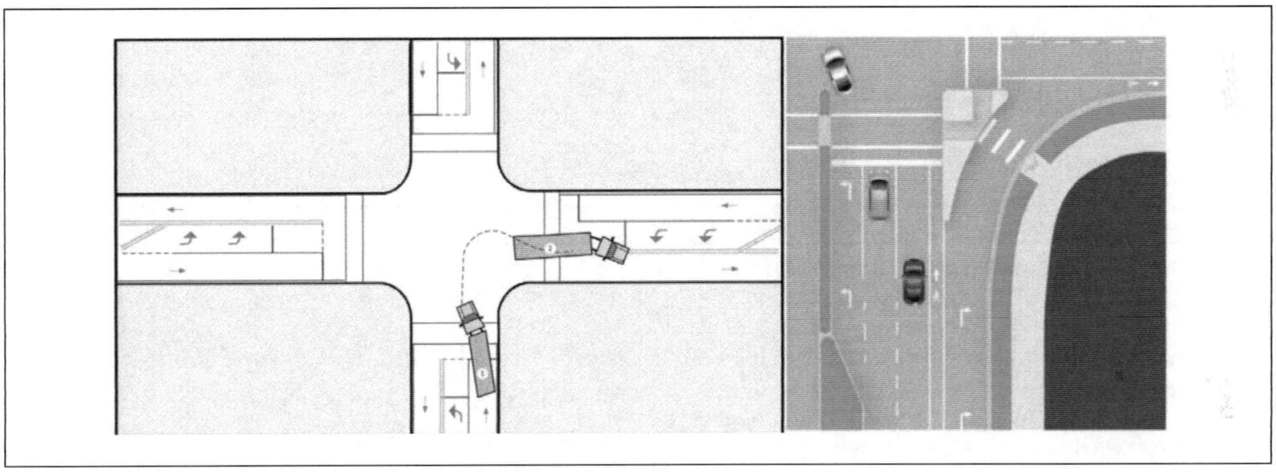

Figure 5. Example of pushed back stop line at an intersection and channelized right-turn.

downtown Dallas. All locations have large numbers of truck crashes that have occurred due to the lack of space for larger vehicles to make right turns. Figure 4 shows an example intersection in Austin, TX, where a crash occurred.

Countermeasures for this crash risk factor are as follows:

- Conduct analyses of turning paths to ensure that large trucks can make turns safely
- Provide clear turning guide lines on the pavement
- Push back the stop lines of the crossing street to provide more space for trucks making right turns (Figure 5, left)
- Use right-turn slip lanes to channelize right turns (Figure 5, right)

Two-Way Frontage Road Design

According to the collision diagrams developed for the hot areas in Midland, TX, seven fatal crashes occurred on the two-way frontage roads in a 5-year span. Figure 6 shows an example crash that occurred at one location indicated in the developed collision diagram. According to the police report, the large truck was traveling westbound and trying to enter the freeway when the passenger car failed to yield to the large truck. In the police report, it was also mentioned that another vehicle continued going through the yield sign and also failed to yield to this truck.

It was found that the design of a two-way frontage road confuses drivers concerning the right-of-way rules, especially at highway entry ramps where the vehicles traveling on two-way frontage roads must make left turns

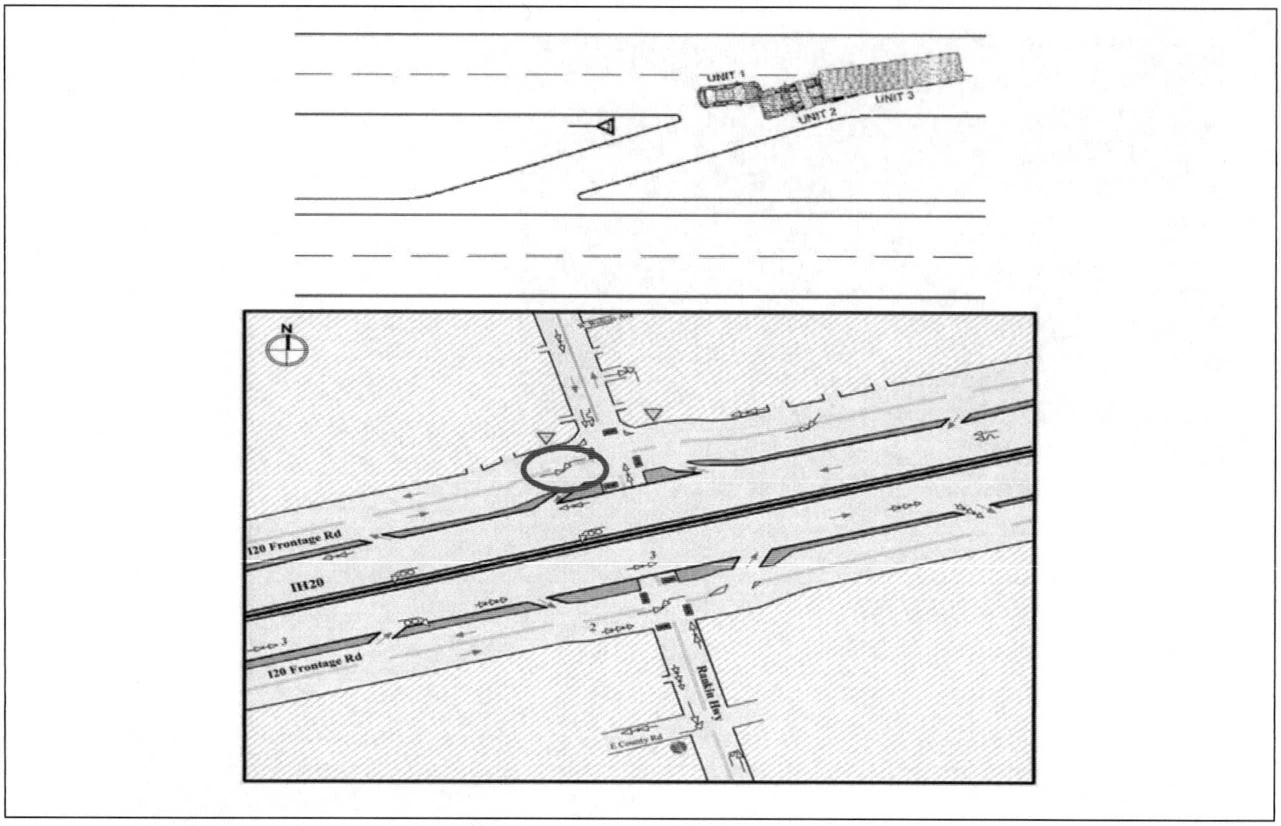

Figure 6. Collision diagram for crash hot area "Midland."

to enter the highway. Although they are left-turning vehicles, the opposing through vehicles still must yield to them because they are entering the highway and have the right of way, which is quite counterintuitive. Therefore, the opposing through vehicles tend not to yield to these left-turning vehicles, resulting in severe head-on crashes. Note that, although there was a "YIELD" sign for the opposing through traffic at the entry ramps, some drivers still failed to see it due to the high speed (speed limit 45 to 50 mph) on the frontage roads.

Crashes caused by this risk factor were usually head-on crashes, which are likely to be serious. Possible countermeasures to reduce the risk of crashes at two-way frontage roads include:

- Convert two-way frontage roads to one-way frontage roads
- Add more traffic signs with actuated flashing lights to the frontage road traffic
- Reduce speed limits on frontage roads around highway entry ramps to reduce the likelihood and severity of crashes
- Educate drivers on who has the right of way and how to operate on a two-way frontage road

Short Merge Distance at Highway Entrances

A direct-taper design of highway entry ramps is used extensively in some areas. When applying a taper ramp, it is critical to provide enough distance for drivers to accelerate and merge into the main lanes. We noticed that there is an entry ramp that has quite a short merge distance. The short distance makes merging difficult for vehicles, especially for large trucks. Figure 7 is a sample collision diagram indicating the location of this type of crash as well as a Google Maps street view of this ramp.

At the location circled in Figure 7, three accidents occurred due to the short merge distance. To reduce angle crashes near highway entrance ramps, the following countermeasures are recommended:

- Extend the merge distance for entry ramps
- Use auxiliary lanes to connect the entry ramps and exit ramps to increase the merge distance

Failure to Control Speed on Highway Exit Ramps

The congestion at frontage intersections causes queues spilling back to the locations of the highway exit ramps,

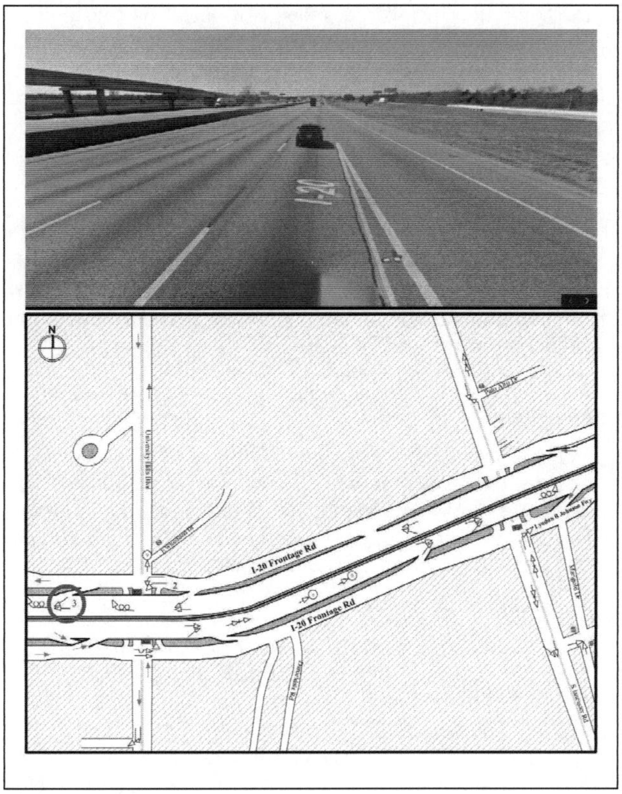

Figure 7. Collision diagram for large truck crash hot area "South Dallas IH20."

which cause vehicles that exit the highway to make a sudden reduction of speed at the end of the exit ramp. Rear-end or angle crashes may occur if drivers are unable to control the speed of their vehicles.

During our analysis, we found that this risk factor was evident at three different identified hot spots: Houston, Midland, and Fort Worth, TX. Figure 8 shows a sample collision diagram for a hot area in Houston, TX. As shown in Figure 8, six crashes occurred at a highway exit ramp location.

Police reports showed that a large truck and a passenger vehicle were exiting the highway. The passenger vehicle was in front of the large truck and stopped due to traffic congestion at the end of the exit ramp. The driver of the large truck failed to control the truck's speed and hit the passenger vehicle.

To reduce rear-end crashes on exit ramps or near the ramp locations, radar speed signs and transverse rubber strips can be utilized.

Conclusions and Recommendations

In this study, historical crash data were used to conduct a risk-analysis study based on collision diagrams. Based on the developed diagrams, the main roadway-related risk factors that contribute to crashes of large trucks were identified. Based on the results of this study, the following key findings and recommendations are presented:

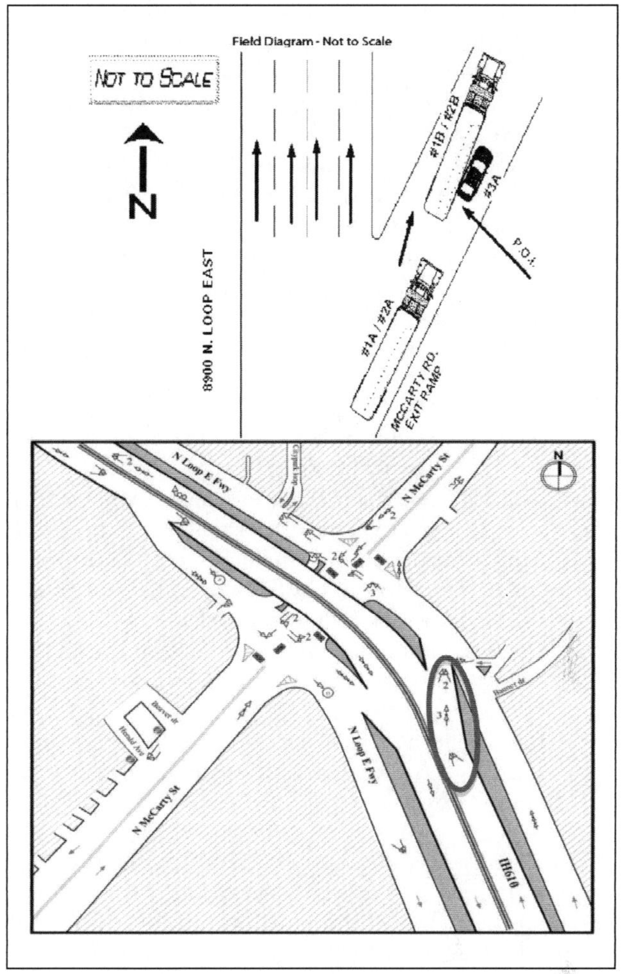

Figure 8. Collision diagram for large truck crash hot area "Houston IH610 @ US90" and sample crash report.

1. The dual left-turn design at intersections with a heavy volume of truck traffic increases the risk of trucks making side-by-side turns. Clear turning pavement markings should be provided at this type of intersection to assist drivers in making safe left turns. In addition, installing a designated signal on each lane is recommended to eliminate any confusion among the drivers.
2. Large trucks require more space to make right turns, and insufficient space can cause crashes when trucks try to make right turns. For the intersections with insufficient space, it is recommended that the stop lines of the crossing street be moved

further back to provide more space for trucks making right turns. In addition, the use of right-turn slip lanes is recommended for intersections that must accommodate a heavy volume of trucks making turns.

3. The design of the two-way frontage road causes drivers to be confused about the right-of-way rules, especially at highway entry ramps. This design should be avoided, and, if it is used, more traffic signs with actuated flashing lights should be installed on the frontage roads near the entry ramps to the highway; further, speed limits on frontage roads should be reduced at these locations.
4. Short merge distances increase the probability of accidents involving large trucks, especially for locations at which there are heavy volumes of trucks. Extending merge distances for highway entry ramps or using auxiliary lanes to connect the entry ramps and exit ramps is recommended for the locations with such problems.
5. Congestion at downstream intersections causes sudden stops at highway exit ramps, and crashes may occur if drivers fail to control the speed of their vehicles. Installing radar speed signs and transverse rumble strips are recommended at such highway exit ramps.

In this study, some unique and local specific risk factors that contribute to crashes of large trucks were identified using the methodology based on collision diagrams. For example, the two-way frontage road design factor in Finding 3 had not been identified previously in research. This indicates that the methodology used in this study can fill gaps in the current truck safety studies by conducting in-depth, locality-specific crash risk analysis. The results of these studies will provide traffic engineers/roadway designers with a better understanding of the risk factors related to the crashes of large trucks. Thus, they will be able to avoid or correct some potential problems on roadways to make them safer for accommodating the movements of large trucks.

Acknowledgments

This paper was prepared based on the research project 0-6911 "Development of Systemic Large Truck Safety Analyses" sponsored by Texas Department of Transportation. We thank the project committee members for their assistance and constructive inputs to the overall research efforts. This research is supported in part by the United States Department of Transportation under grant no. 69A3551747133 and the U.S. Department of Homeland Security under grant no. 2014-ST-062-000057-02.

Author Contributions

The authors confirm contribution to the paper as follows: study conception and design: YQ, QZ; data collection: TG, QZ; analysis and interpretation of results: QZ, MA, TG; draft manuscript preparation: QZ, TG, YQ. All authors reviewed the results and approved the final version of the manuscript.

References

1. Federal Motor Carrier Safety Administration. *Report to Congress on the Large Truck Crash Causation Study*. U.S. Department of Transportation, Washington, D.C., 2006. https://www.fmcsa.dot.gov/safety/research-and-analysis/report-congress-large-truck-crash-causation-study. Accessed August 8, 2018.
2. Garber, N. J., and A. A. Ehrhart. *The Effect of Speed, Flow, and Geometric Characteristics on Crash Rates for Different Types of Virginia Highways*. Final Report. Publication FHWA/VTRC 00-R15. Virginia Transportation Research Council, Charlottesville, VA, 2000.
3. Jerry, G. J., and K. R. Agent. *Heavy Truck Involvement in Traffic Crashes and Related Countermeasures*. Report No. KTC-99-20. Kentucky Transportation Center, Lexington, KY, 1999.
4. Garber, N. J., and S. Joshua. Characteristics of Large Truck Crashes in Virginia. *Transportation Quarterly*, Vol. 43, No. 1, 1989, pp. 123–128.
5. Burke, P., J. Carson, and R. Bylsma. *Characterizing Commercial Vehicle Safety in Rural Montana*. Motor Carrier Services Division, Montana Department of Transportation, Helena, MT, 2002.
6. Ragin, C. C. *Fuzzy-Set Social Science*. The University of Chicago Press, Chicago, IL, 2000.
7. Cooner, S. A., S. E. Ranft, Y. K. Rathod, Y. Qi, L. Yu, L. Wang, and S. Chen. *Development of Guidelines for Triple Left and Dual Right-Turn Lanes*. Technical Report, No. 0-6112-1. Texas A&M Transportation Institute, College Station, TX, 2011.

The Standing Committee on Truck and Bus Safety (ANB70) peer-reviewed this paper (18-03220).

The contents of this paper reflect the authors' views, who are responsible for the facts and accuracy of the data presented.

TRR
JOURNAL OF THE TRANSPORTATION RESEARCH BOARD

Article

Transportation Research Record
2018, Vol. 2672(34) 29–38

DOI: 10.1177/0361198118792132
journals.sagepub.com/home/trr
SAGE

Challenges in Hazard Detection for Commercial Motor Vehicle Drivers: A Driving Simulator Training Study

George D. Park[1], Jeffrey S. Hickman[2], Sean P. Pitoniak[1], and Theodore J. Rosenthal[1]

Abstract

There are few research studies that have assessed the hazard detection (HD) challenges of experienced, skilled drivers. This paper presents the pilot study results of an HD training program, Commercial Driver Assessment and Training System (CDATS), developed for commercial motor vehicle (CMV) drivers. The training involved simulation-based videos that highlight potential vehicle, pedestrian, and visually hidden hazards during unprotected intersection maneuvers: left-turn, right-turn, and straight-thru. Low-fidelity driving simulations were designed to reinforce video lessons and increase the expectation and detection of potential hazards. Training efficacy was assessed using a driving simulator HD task presented to short-haul CMV drivers ($N = 16$) at pre-/post-assignment to CDATS training or a control condition. Results suggested that CMV drivers had highest HD task accuracy (% correct) for vehicle hazards and lowest accuracy for visually hidden hazards. Performance for pedestrian hazards was dependent on the intersection maneuver, with performance decreases for left-turn, right-turn and straight-thru intersections, respectively. Post-test results suggested overall HD improvement in the CDATS training group with no change in the control group; however, there was no significant change in relative hazard type by intersection maneuver pattern. Similar to prior research, experienced CMV drivers may also exhibit challenges in HD, particularly for visually hidden and visible pedestrian hazards, and may benefit from HD training.

A driving hazard can be defined as anything (e.g., vehicle, pedestrian, object) or situation (e.g., visual obscurity, roadway condition) that may result in a collision, requiring anticipatory or evasive action by the driver (e.g., brake pedal covering, speed reduction, or steering). Historically considered a novice driver (aged 16–24) or older driver (aged 65 +) problem due to their higher crash rates (*1*), a driver's lack of hazard detection (HD) ability can be associated with increased crash risk (*2, 3*). Unlike older drivers, who may be dealing with age-related deficits, novice drivers are thought to exhibit poorer HD skills due to inexperience, with the idea that HD skills eventually develop as drivers gain maturity and exposure. However, middle-aged drivers, though still outperforming novices, can exhibit suboptimal HD skills despite decades of driving experience (*4*). These findings may explain the crash data found for those individuals that arguably drive the greatest number of miles, commercial motor vehicle (CMV) drivers.

Motor vehicle crashes remain the leading cause of death among U.S. workers in an industry that includes more than 539,000 active motor carriers and 5.6 million commercial drivers (*5*) and costs employers US$60 billion annually in medical care, legal expenses, property damage, and lost productivity (*6*). Studies such as the Large Truck Crash Causation Study (*7*) have found crashes are overwhelmingly caused by human error (89% human, 8% vehicle, and 3% roadway environment) with the most common human error types being: 38% driver decision error (e.g., too fast for conditions), 29% driver recognition failure (e.g., inattention), and 12% driver physical failure (e.g., asleep-at-the-wheel). Therefore, driver training that focuses on HD skills to initiate *pre-event* behaviors prior to crash events (e.g., speed, headway, gap acceptance, braking) may be the most effective means of preventing crashes. However, little is known regarding the specific types of hazards and situations that CMV drivers find most challenging and the best methods

[1]Systems Technology, Inc., Hawthorne, CA
[2]Virginia Tech Transportation Institute, Blacksburg, VA

Corresponding Author:
Address correspondence to George D. Park:
georgepark@systemstech.com

for improving HD skills. In other words, "How do you get experienced, skilled drivers to begin attending to different hazardous areas to improve their overall situational awareness?"

The Salience, Effort, Expectancy, and Value (SEEV) Model

A useful framework for understanding why a driver may not be properly attending to areas in the environment or gathering all the information for making decisions is the salience, effort, expectancy, and value (SEEV) model of visual information sampling (*8*). Utilized in modeling the visual scanning behavior of operators in aviation (*9*, *10*) and driving (*11*), the SEEV model can potentially explain the visual attention differences between novices and experts, and the effects of driver state, task workload, and environmental contexts.

According to the SEEV model, the probability of a skilled driver selectively attending to or sampling sources of information from an area of interest (AOI), is influenced by the combined factors of SEEV (salience, effort, expectancy, value). *Salience* refers to an AOI's ability to capture or grab attention (e.g., flashing lights). *Effort* refers to the amount of resources needed to access the information at the AOI including physical (e.g., visual angles for head-eye movements) and mental workloads (e.g., concurrent tasks). Increases in salience lead to increases in AOI attention, whereas increases in effort lead to an inhibitory effect in AOI attention. Both salience and effort are bottom-up or environmentally driven forces, often out of the control of the driver. *Expectancy* refers to the frequency or probability of seeing relevant information in an AOI. *Value* refers to the importance—or cost of missing—the information from an AOI to the task(s) at hand. Increases in expectancy and value lead to increases in AOI attention. Both factors are top-down or knowledge driven forces, related to driver experience and situational awareness.

Applied to driver HD training, the SEEV model guides how to improve a proficient, skilled driver into a better driver or, alternatively, maintain a more appropriate state of vigilance for high value hazards while driving. This can potentially be achieved by increasing or maintaining the *expectancy* and *value* factors for hazards. For example, consider a situation where a large truck is parked on the right side of the roadway at a pedestrian crossing. On approach, novice drivers often fail to look toward the truck even though their visual field will be vulnerable to crossing pedestrians, hidden from the right (*12*). Their expectation of a pedestrian hazard risk is low. As novice drivers gain roadway experience, their expectation of a pedestrian risk at that crossing appropriately increases, making it more likely that they look toward the truck. However, experienced drivers may *decrease* their expectation and value for the *same* hazardous event if their continued experience proves that no pedestrians ever cross. As such, experienced drivers may eventually stop looking toward that hazardous truck and be unprepared for pedestrian crosses. The rarity of some hazards during everyday driving may ill-prepare experienced drivers when those hazards do inevitably occur.

CMV Driver Training using Simulation Technology

Proactive, traffic safety programs presented to employee drivers at initial training and throughout a driver's career are potentially one of the best ways to control costs from workplace vehicle crashes. Reactive training methods (e.g., evasive maneuvering/braking) are important, but not as effective as preventing risky crash situations in the first place. For an HD training program, driving simulation technology provides an attractive solution. Simulation allows sensory and motor skills and behavioral tendencies to be trained and assessed. The many advantages include: safety, scenario versatility, standardization, thoroughness, repeatability, improved perspectives, sophisticated measurement, and efficiency (*13*).

Effective simulations can be expensive to develop, particularly when the needs of the industry are so broad. CMV operations can be small—a single vehicle, owned and operated by a single individual; or large—a corporation that owns tens of thousands of vehicles. Operations can differ in terms of what they carry (e.g., freight, passenger, hazardous materials), where they operate (e.g., locally, regionally, cross country), when they operate (day, night, seasonal), and the number of destination points (e.g., truckload or less-than-truckload) (*14*).

Simulation-based training can also be difficult to deploy in a cost-effective manner. For larger carriers with thousands of existing drivers and high turnover rates, deploying simulation-based training programs can be unwieldly to manage, time consuming, and impractical to implement fleet-wide regardless of the training's efficacy. Similarly, for smaller carriers the costs may be too prohibitive, particularly if specialized simulation hardware (e.g., driver display/controls, seating, etc.) and support structures (e.g., trained instructors, classroom space) are required. Even when the above challenges are met, the resulting performance data are not easily interpretable. With the discrepancy in simulation hardware and scenario design, normative data for simulation performance remains elusive in driver assessment and training applications.

CDATS Training Approach

The Commercial Driver Assessment and Training System (CDATS) is a HD training program developed for CMV drivers. At present, it focuses on short-haul, CMV drivers (e.g., package delivery). It employs a combination of multimedia materials, simulation-based videos, and low-fidelity driving simulation technology to provide a cost-effective and efficient means of improving the HD skills of drivers for everyday driving environments.

The CDATS HD training approach involves several elements designed to increase the SEEV model factors of expectation and value for potential hazards. It is also designed to convey the breadth of potential hazards to spread attention across the driver's visual field, as opposed to focusing on specific critical hazards that experienced drivers likely already respond to. This may in turn lead to greater situational awareness and drivers responding more quickly with appropriate pre-event behaviors.

Focus and Breakdown of High-Risk Maneuvers. CDATS currently focuses training on high-risk vehicle maneuvers where traffic and pedestrians intersect. Prior naturalistic driving studies for short-haul trucks have indicated that the most common incident types involve lane change without sufficient gap, roadway entrance without sufficient clearance, and left-turn without clearance (*15*). These situations involve a high number of AOIs on which drivers must distribute their attention across the driving scene. The maneuvers addressed by CDATS are left-turns, right-turns, and straight-thru at unprotected intersections (i.e., no dedicated turn signal arrow/lane). Each of these maneuvers is further broken-down into three zones where hazards may appear:

- Pre-maneuver: preparation period for the maneuver, e.g., driver changes lanes to make a left-turn ~200 ft prior to the intersection. Hazards such as pedestrian crossings may be similar for different maneuvers depending on driver vehicle positioning.
- Maneuver: execution period for maneuver, e.g., driver executes a left-turn across oncoming traffic and pedestrians at the crosswalk. Hazards are unique to maneuver.
- Post-maneuver: stabilization period immediately after the maneuver, e.g., after a wide right-turn, driver moves back to far-right lane. Like pre-maneuver hazards, but locations are subject to the maneuver. Requires most anticipation and situational awareness as it occurs quickly after maneuver execution and may involve overlap with the next pre-maneuver period.

Video and Simulation Techniques to Visualize the Intersection scene. For each maneuver type, CDATS provides videos embedded with text, audio narration, and driving simulation scenes to highlight hazards and practice real-time hazard perception. Overhead camera angles are utilized to overcome typical driver field-of-view (FOV) limitations and improved visualization of the traffic scene, especially for hazards involving visual FOV obstructions (e.g., large truck obstructing the view of a pedestrian).

As shown in Figure 1 for an unprotected left-turn maneuver, a driving simulation video starts from a typical driver FOV. As the video progresses, it pauses at a pre-maneuver zone and asks, "Where are the potential hazards prior to the intersection?" (Figure 1*a*). The camera angle then shifts to an overhead view and all potential hazards are explicitly highlighted (verbally and graphically) in a serial fashion (Figure 1*b*). The camera angle then returns to the typical driver FOV and the video proceeds to the actual intersection where the video pauses again and asks, "Where are the potential hazards during and immediately after your left turn?" (Figure 1*c*). As before, the camera angle shifts to an overhead view and all potential maneuver and post-maneuver hazards are highlighted (Figure 1*d*). The camera angle returns and the simulation continues. The entire intersection scene is then repeated in the normal driver FOV to allow students to practice detecting and anticipating hazards in real time.

Reinforcement through Driving Simulation. Video lessons are reinforced using the same driving scenarios presented on a single-screen, low-fidelity, desktop simulator. Potential hazards (previously static for highlighting purposes in the video lessons) are now activated to cross the driver's path, allowing students to practice the appropriate situational responses in real time (e.g., speed reduction, brake covering, delayed execution) to avoid collisions. In these scenarios, the rare occurring hazards are just as likely, if not more so, to be presented to drivers; thus, potentially increasing the *expectation* factor of specific hazards.

Simulation drives are performed by the trainee using 1) a normal driver FOV with minimal cabin instrumentation and two sideview-mirrors, and 2) an elevated driver FOV that retains the sideview-mirrors but presents a wider FOV of the scene with the roof of the driver's vehicle at the bottom of the screen (Figure 2). The elevated driver FOV configuration was designed to correspond to the video lessons and overcome the limited driver FOVs in typical single-screen driving simulations where frequent horizontal scene panning is required. Such driving configurations may provide useful training methods for low-fidelity simulation training in which situational awareness training is emphasized over vehicle control.

Current Study

To assess the training potential of the CDATS program, a pilot-training study was performed using a sample of experienced, short-haul CMV drivers. A driving simulator HD task was developed that was composed of a variety of potential hazard types (e.g., vehicle, pedestrian,

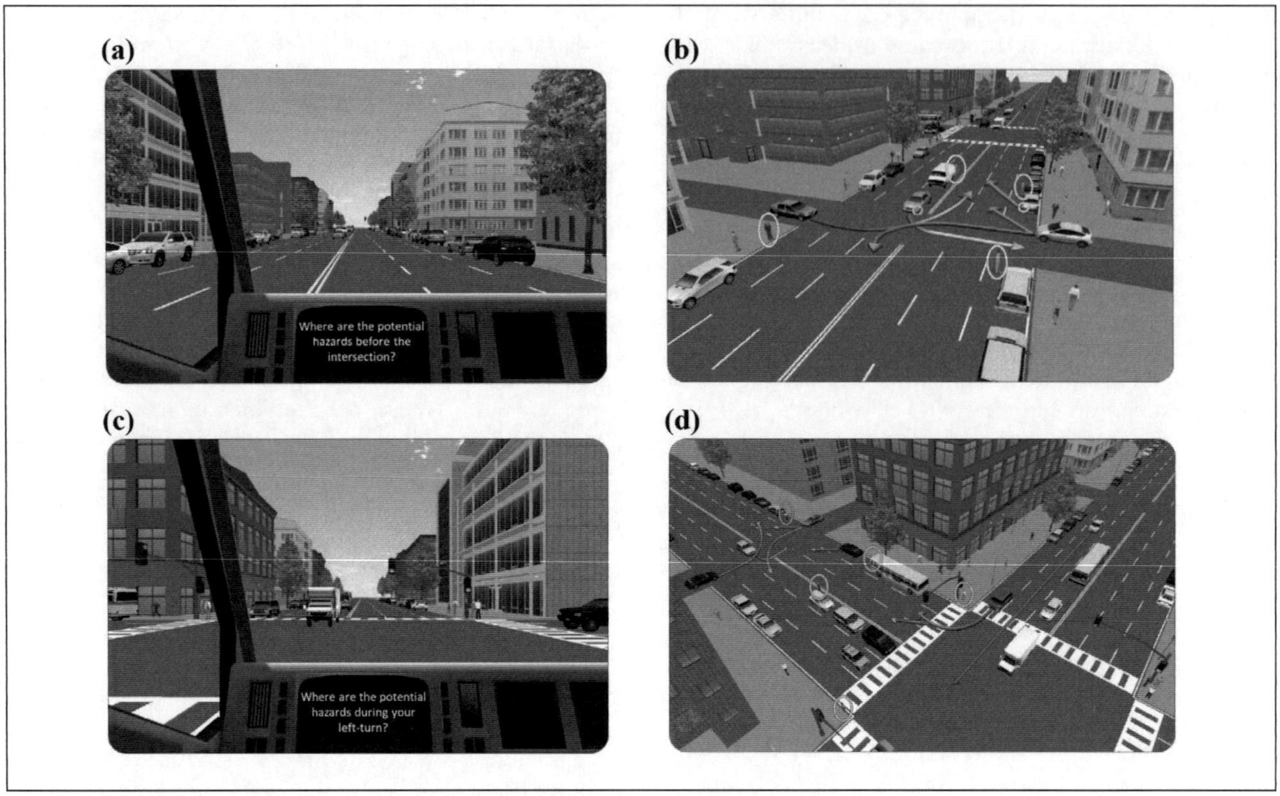

Figure 1. CDATS training video scenes using embedded simulation for an unprotected left-turn intersection maneuver. Scenes pause (*a*) and (*c*) and camera angles shift to highlight hazards prior to intersection (*b*) and during/after the left-turn maneuver (*d*).

and visually hidden hazards) during three unprotected, intersection maneuvers: left-turn, right-turn, and straight-thru. A potential hazard was defined as anything or any situation that may result in a collision, requiring an evasive action such as slowing down, braking, or steering. The HD task was presented to participants before random assignment to either the CDATS training or control group (pre-test) and 1-week later (post-test). Control group participants were provided with a series of hazard recognition and response scenarios utilized in prior novice driver training studies (*16*).

CMV driver HD performance and CDATS training efficacy was assessed using HD task accuracy (arcsine-square root transformation of % correct of total potential hazards). Although CDATS trained participants were expected to have better HD task accuracy than the control group, the overall pattern of hazard perception deficiencies (i.e., hazard types by intersection maneuvers) for CMV drivers may prove useful for researchers and future interventions.

Methods

Participants

Sixteen subjects (15 male, aged 24–57) participated in this study. All subjects were professional drivers (e.g., short-haul, package delivery) with an average of 15 years of work experience (range: 3–33 years, $SD = 11$ years) with an average of 73-mi driven per shift ($Mdn = 55$ mi, $SD = 60$ mi). Recruited from advertisements (Internet and flyers), subjects were compensated US$75 per session completed, two sessions in total.

Driving Simulator

The driving simulator system (Figure 3) was generated by STISIM Drive® (Build 3.14.04) and processed by a Dell Precision Tower 5810 (Win 10 OS, 3.10 GHz, 8 GB RAM, NVidia GeForce GTX 1060). The Logitech Driving Force GT steering wheel and pedals (900° rotation, automatic transmission) was used for driver controls and recording the HD task button responses.

For the pre/post-test drives, a 150° horizontal FOV of the driving scene was provided using three monitors (22 in. diagonal, 1,680 x 1,050 pixels). For the CDATS training and control group drives, a 60° horizontal FOV was provided using only the center monitor. The elevated FOV setting was established by situating the driver's eye/camera position in the virtual scene, relative to vehicle model center: 57-ft vertical, –74-ft longitudinal, and −20° downward pitch.

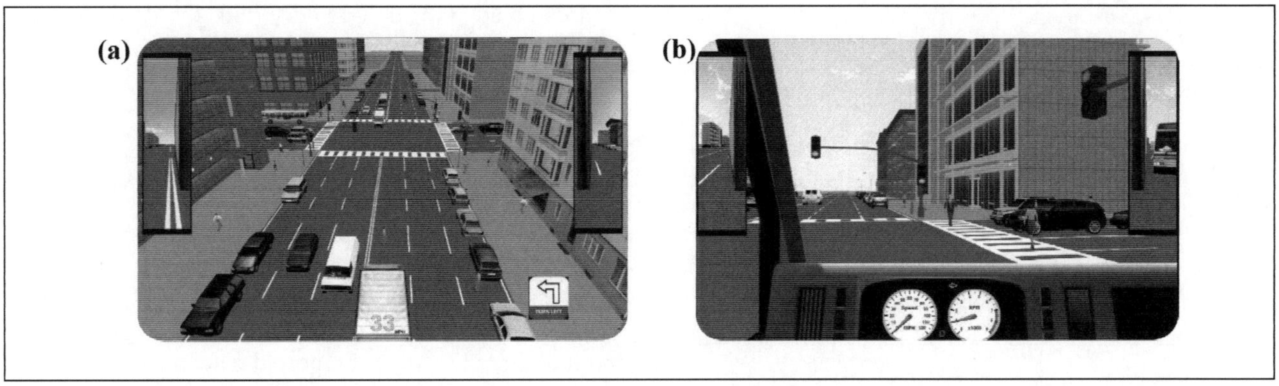

Figure 2. CDATS driving simulation scenes: (*a*) elevated driver FOV, white truck at center bottom is driver's vehicle with speedometer and turn instructions displayed, and (*b*) normal driver FOV as driver's vehicle is making a right-turn.

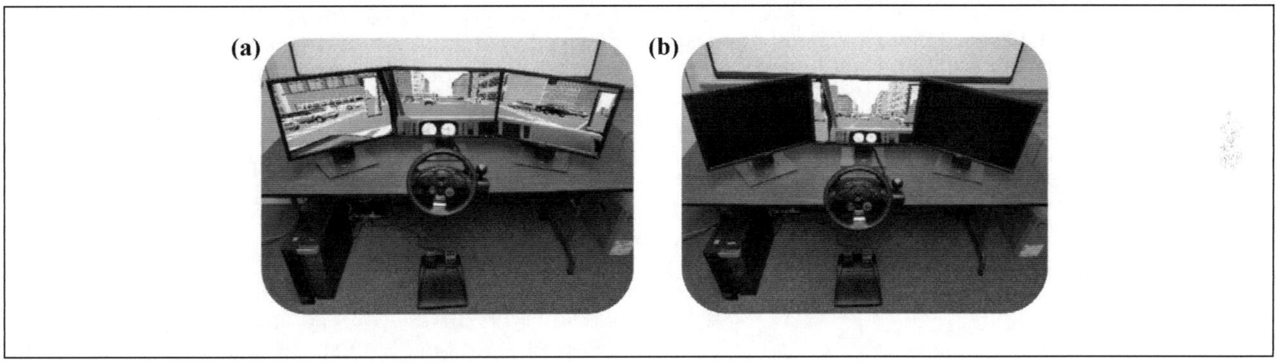

Figure 3. (*a*) Driving simulator configurations for pre/post-tests, and (*b*) CDATS/control group training.

Pre/Post-Test Drives

Nine intersections (three left-turns, three right-turns, three straight-thru) spaced ~2,000-ft apart were provided for a total driving distance of 8,900 ft (3.6 mi); drive time: ~8 min. Each intersection scene had the following design: ~1,100 ft in driving length, six lanes across (12 ft each), double-yellow centerline, unprotected green signal lights, intersection crosswalks, two-lane side streets ~200 ft before/after and left/right of the intersection, and urban style building models. Minimal ambient, oncoming traffic was provided between intersections.

Hazard Types and Placement. All potential hazards were static in scene position. Vehicle hazards involved waiting vehicles signaling their intentions to cross or change lanes in front of the driver. Pedestrian hazards involved pedestrians standing 10-ft off-sidewalk, facing road-center or on-sidewalk at one of the intersection crosswalk corners. Hidden hazards involved a large vehicle (e.g., delivery truck or bus) obstructing the view of either traffic or pedestrians.

For each intersection type (e.g., left-turn), a single pedestrian or vehicle hazard was positioned: (i) immediately prior to the intersection (i.e., within 200 ft), (ii) within the immediate area of the intersection maneuver, or (iii) immediately after the intersection maneuver. A hidden hazard was positioned once during the pre-maneuver or maneuver sections. A third of the sections did not contain any hazards. Therefore, a single test drive resulted in 12 visible hazards (six pedestrians, six vehicles) and six hidden hazards (half large truck, half transit bus). Half the hazards were positioned on either the right or left side of the driver's FOV. Intersection maneuvers and hazards were presented in a random order. For the post-test drive, building models were rearranged while maintaining the same road design and a counterbalanced version of the intersection types and hazards were presented.

Hazard Detection Task Procedure. When a participant detected a potential hazard, they were instructed to press the HD button located on the steering wheel as quickly as possible. When pressed, the HD button paused and

Table 1. CDATS Training Curriculum

Training items	Time (min)
Training Videos	
CDATS Introduction, Motivation, and Goals	4
Unprotected Left-Turn Intersections	4
Unprotected Right-Turn Intersections	4
Unprotected Straight-Thru Intersections	4
Driving Simulator Training	
Elevated FOV Driving Simulation	5
Normal FOV Driving Simulation 1	5
Normal FOV Driving Simulation 2	5

blacked-out the simulation scene. The participant then verbalized to the experimenter what and where the potential hazard was. When finished speaking, the participant re-pressed the button to unpause and continue the test drive. Performance feedback was only provided at study completion. Vehicle speed maxed at 40 mph to minimize hazard approach timing differences.

Additional participant instructions included: (i) drive as if making normal deliveries in your work vehicle, (ii) maintain 40 mph when safe to do so, (iii) pay attention to any turn instructions, (iv) there is no right or wrong way to drive and you will not be judged on how you drive, (v) it is ok if you do not detect any hazards as not every section will contain a hazard, and (vi) point out more than one hazard during a pause if noticed.

CDATS Training Group Curriculum

As shown in Table 1, the CDATS training curriculum comprised of a series of videos that included a training introduction (including motivation and goals), followed by hazard training videos for each intersection maneuver type. This was followed by driving simulator training that used an elevated FOV configuration and then normal FOV training drives. Each of the training drives exposed drivers to the three intersection maneuver types with hazards activated to collide or near collide with the driver's vehicle if action was not taken. Total training time: 31 min.

Control Group Scenarios

For the control group, a total of five driving scenarios (total drivetime: 30 min) were selected to improve vehicle handling skills and expose participants to typical hazard detection/response drives as used in prior novice driver training studies (*16*). The first scenario was a raceway drive where the goal was to pass as many race cars as possible. The remaining four scenarios involved a variety of driving speed and building environments (e.g., residential, urban, and rural), where pedestrian and vehicle hazards could potentially collide with the driver if they did not take evasive action.

Additional Test Materials

Questionnaires were developed to collect basic demographic information, current health status, driving history (e.g., professional experience, average mileage/shift, past crash/citation within 2 years), and system usability. The Simulator Sickness Questionnaire (SSQ) was used to measure any simulator sickness symptomology (*17*). SSQ and usability data are not reported here.

Procedure

Data was collected using two separate sessions. At Session 1, the participant began by completing the informed consent, questionnaires, and SSQ baseline. The participant was then introduced to the driving simulator and HD task using the practice drive. When comfortable with the simulator and tasks, the participant was given the pre-test drive. After the drive, participants completed a post-test drive SSQ and usability survey.

Training group participants were then provided the CDATS training curriculum. Control group participants were provided the control group scenarios. Once training or simulator driving was complete, participants completed an additional SSQ survey and were excused with compensation (US$75). Total time for Session 1 was approximately 1.5 h.

Participants returned 1-week later for Session 2 to perform the post-test drive. Prior to the test drive, participants were asked if any changes in health had occurred in the last week and a baseline SSQ was collected. The practice drive was provided to refresh participants' memories on the driving simulator and a similar test procedure was used for the post-test drive. When finished, a post-test drive survey was provided. For the training group, an exit interview was conducted. For the control group, the option of experiencing the CDATS training materials was provided. Participants were then debriefed, compensated (US$75), and excused. Total session time was approximately 1 h.

Results

A 2 (group: training, control) × 2 (test: pre, post) × 3 (hazard type: vehicle, pedestrian, hidden) × 3 (intersection maneuver: left, right, straight) mixed analysis of covariance was performed for HD task accuracy (arcsine-square root transformed, % correct), $N = 16$, $\alpha = .05$. The covariate factor was driving experience (years driving professionally).

Table 2. HD Task Accuracy, Mean (*SD*) % Correct for Hazard Types by Intersection Maneuver

	Intersection maneuver		
Hazard type	Left turn	Right turn	Straight thru
Vehicle	92 (15)	78 (27)	92 (15)
Pedestrian	73 (27)	52 (17)	30 (23)
Hidden	9 (22)	20 (23)	19 (30)

Note: Values untransformed and unadjusted for covariate.

A significant hazard type × intersection maneuver interaction, $F(4,52) = 6.39$, $p < .001$, $\eta^2_p = .329$, suggested that participants missed hazards differently depending on the types of maneuvers performed at an intersection. A main effect for hazard type, $F(2,26) = 29.30$, $p < .001$, $\eta^2_p = .693$, was also found. The covariate, driving experience, was not related to HD task accuracy ($p = .982$). Transformed means reported in text with ± standard error and adjusted for a covariate value of 15.3 years driving professionally. To aid interpretation, Table 2 provides untransformed and uncovariate adjusted, % correct means (*SD*) for each hazard type across intersection maneuvers.

For each intersection maneuver, simple main effects were assessed followed by Bonferroni pairwise comparisons. As shown in Figure 4, at intersections where drivers made unprotected left-turns, $F(2,28) = 21.97$, $p < .001$, mean HD task accuracy for vehicle hazards was high (1.45 ± .06), but not statistically different from pedestrian hazards (1.15 ± .11), $p = .102$. Mean accuracy for hidden hazards during left-turns (.15 ± .09) was very low compared with pedestrians ($p < .001$) and vehicles ($p < .001$).

At intersections where drivers made unprotected right-turns, $F(2,28) = 7.61$, $p = .002$, mean HD task accuracy for vehicle hazards (1.23 ± .12) was higher compared with pedestrian hazards (.81 ± .07), $p = .005$. Mean accuracy for hidden hazards (.34 ± .08) was the lowest compared with vehicles ($p < .001$) and pedestrian hazards ($p = .003$).

At intersections where drivers continued straight-thru, $F(2,28) = 13.94$, $p < .001$, mean HD task accuracy was higher for vehicle hazards (1.45 ± .06) compared with both pedestrian (.47 ± .09), $p < .001$, and hidden (.30 ± .11), $p < .001$, hazards. No difference found between pedestrians and hidden hazards ($p = .328$).

A significant group × test interaction was found, $F(1,13) = 7.97$, $p = .014$, $\eta^2_p = .380$. Subsequent *t*-tests confirmed no difference in mean HD task accuracy between the training (.72 ± .08) and control group (.73 ± .08) at pre-test ($p = .913$). However, at post-test, the training group scored on average better (1.03 ± .07) than the control group (.77 ± .07), $t(14) = 2.55$, $p = .023$

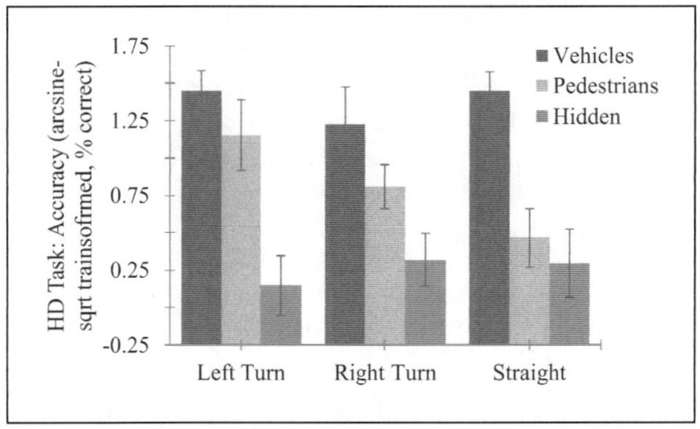

Figure 4. Mean HD task accuracy by hazard type across intersection maneuvers with 95% CIs.

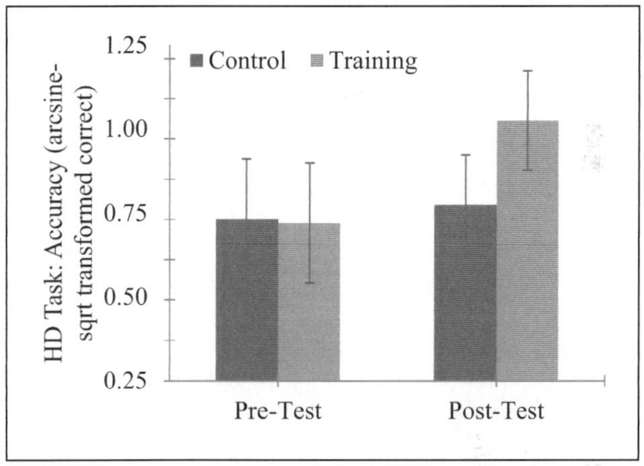

Figure 5. Mean HD task accuracy at pre/post-tests by group with 95% CIs.

(Figure 5). To aid interpretation, pre-transformed values (mean ± SE for % correct) for the control group at pre- and post-test were 47 ± 5% and 49 ± 4%, respectively. For the trained group, pre-transformed values at pre- and post-test were 46 ± 5% and 65 ± 4%, respectively. No four-way interaction (group × test × hazard type × intersection maneuver) was found, $F = 1.75$, $p = .154$, suggesting that no significant change in hazard × intersection maneuver pattern was observed between the training and control group at post-test.

Discussion

To assess the HD training potential of CDATS, a pilot-training study was performed using a sample of experienced, short-haul CMV drivers, $N = 16$. HD task accuracy (arcsine-square root transformed, % correct) was measured using a wide-FOV driving simulator and HD task composed of potential hazards (e.g., vehicle, pedestrian, and visually hidden hazards) during unprotected intersection maneuvers (left-turn, right-turn, and straight-thru). The HD task was presented to participants before random assignment to either the CDATS training or control group (pre-test) and 1-week later (post-test).

Hazard Type by Intersection Maneuver

CMV drivers had the least difficulty detecting potential vehicle hazards that were visible and signaling their intentions. Relative to pedestrian hazards, vehicles were larger in size and had turn signal light motions to aid attentional capture. This likely contributed significantly to higher HD task accuracy scores. Vehicle collisions may also represent more of a perceived threat to the driver's personal safety compared to pedestrian collisions. Thus, drivers were perhaps prioritizing the detection of vehicle hazards over pedestrian or hidden hazards.

CMV drivers had the most difficulty detecting the hidden potential hazards, regardless of the intersection maneuver. Although the obstructing objects (e.g., bus) were larger in size than the visible vehicle hazards, drivers did not view them as a visibility threat, let alone expect pedestrians or vehicles to be located behind them. Visibility-based hazards present a clear challenge for even experienced, professional drivers.

CMV drivers detected visible pedestrian hazards differently depending on the intersection maneuver performed. At unprotected left-turns, pedestrians were detected as well as vehicle hazards. At unprotected right-turns, pedestrians were detected less but better than hidden hazards. This difference between left- and right-turns may have been due to vehicle positioning differences while waiting to execute the turns. Drivers during unprotected left-turns could be positioned more center of the intersection, providing a potentially better visual FOV for hazard scanning to the upcoming left roadway. During right-turns, the driver's vehicle usually remains short of the intersection crosswalk, limiting the FOV to the right.

At unprotected straight-thru intersections, pedestrians were detected no better than hidden hazards. The drop in HD accuracy for pedestrians during straight-thru intersection maneuvers was potentially due to the timing demands of the maneuver. At turning intersections, drivers did not necessarily have the right-of-way and had time to wait and scan the environment prior to turning. In contrast, drivers during straight-thru intersections had the green light and right-of-way, providing drivers with less time to detect potential threats. Future studies imposing time restraints on intersection turning maneuvers may reveal pedestrian HD deficits like straight-thru maneuvers. It is also possible that drivers during the straight-thru intersections did not expect pedestrians to cross—visible or otherwise. Interestingly, they did not assume the same from other vehicle drivers, as evident in their high detection of vehicle hazards.

Training Effects

The CDATS program has the potential to improve hazard perception skills for CMV drivers during unprotected intersection maneuvers. At pre-test, no group differences were observed for HD task performance. At post-test, those trained with the CDATS program had improved their HD task performance compared with the control group. Differences between groups at post-test could not be attributed to one group being exposed to more driving simulator scenarios. Despite driving scenarios previously utilized in novice driver training studies in which collision avoidance was required, control group drivers did not appear to gain any additional hazard perception skills. This suggests that experienced, professional drivers may benefit more from cognitive-based training programs for HD. However, novice drivers or other driving populations may also potentially benefit.

Although an overall improvement in HD task accuracy was found for the CDATS group, gains in HD task performance were relatively modest. No significant four-way interaction was found, suggesting that after training, there was no strong difference from the control group in the kinds of hazards that were detected depending on the intersection maneuver. It is possible that the driving simulator HD task was too difficult or that improvements in CDATS training are warranted. The training videos were passive in nature and did not provide any performance feedback. Therefore, it is unclear whether lessons were fully understood, or practiced enough to translate into better simulator performance. Future CDATS designs will require a method of assessing the training process of drivers. At the very least, the HD task method used appeared suited to assessing HD skills of CMV drivers and sensitive to training effects.

Conclusions

In accordance to prior research (*4*), experienced professional drivers can also exhibit poor HD—particularly for visually hidden and visible pedestrian hazards—and may benefit from HD training. Although prior driver training studies have used birds-eye-view images of the roadway (*18*), real-time traffic videos (*19*, *20*), and driving simulation scenarios requiring collision avoidance (*16*, *21*), CDATS is one of the few programs attempting to combine the advantages of the aforementioned programs (see McDonald et al. [*22*] for a review). Our method of using simulation footage lends itself to more efficient curriculum development, scene control, and design flexibility.

It remains to be seen whether the CDATS program translates to reductions in real-world crash risks and rates. Our sample and training system was also specific to short-haul CMV drivers (e.g., package delivery). Different HD training challenges may also appear for long-haul CMV drivers. In addition, the types of hazards trained and assessed were limited to adult-sized pedestrians, passenger vehicles, and visual obstructions based on either a large truck or bus. Since training effects were assessed only after a period of 1 week, the long-term effects of CDATS HD training are also unknown. Questions remain regarding when in a CMV driver's career these types of training interventions are most effective and how a skills maintenance schedule may mitigate future HD deficits from appearing over time.

Future work involves a larger training validation study with instrumented vehicle tests and eye-tracking metrics. Program development involves the introduction of different intersection maneuvers (e.g., unprotected t-intersections) and hazard types (e.g., varying pedestrian sizes, cyclists, motorcycles, and roadway-based visual obstructions). Plans are also intended to develop an interactive and performance feedback version of the video training lessons, which may provide an additional means of assessing HD skills (when no feedback is provided) and training fidelity checks to ensure lessons goals are achieved.

Acknowledgments

This research and development was funded by a Small Business Innovative Research grant from the National Institute for Occupational Safety and Health. Systems Technology, Inc. is a privately owned company that distributes the STISIM Drive driving simulator used in this study.

Author Contributions

The authors confirm contribution to the paper as follows: study conception and design: GP, JH; data collection: GP, SP, TR; analysis and interpretation of results: GP, JH; draft manuscript preparation: GP. All authors reviewed the results and approved the final version of the manuscript.

References

1. Insurance Institute for Highway Safety Highway Loss Data Institute. *Older Drivers*, Retrieved January 4, 2010, http://www.iihs.org/iihs/topics/t/older-drivers/fatalityfacts/older-people/2015.
2. Boufous, S., R. Ivers, T. Senserrick, and M. Stevenson. Attempts at the Practical On-Road Driving Test and the Hazard Perception Test and the Risk of Traffic Crashes in Young Drivers. *Traffic Injury Prevention*, Vol. 12, No. 5, 2011, pp. 475–482.
3. Horswill, M. S., K. J. Anstey, C. G. Hatherly, and J. M. Wood. The Crash Involvement of Older Drivers is Associated with their Hazard Perception Latencies. *Journal of the International Neuropsychological Society*, Vol. 16, No. 5, 2010, pp. 939–944.
4. Horswill, M. S., K. Taylor, S. Newnam, M. Wetton, and A. Hill. Even Highly Experienced Drivers Benefit from a Brief Hazard Perception Training Intervention. *Accident Analysis & Prevention*, Vol. 52, 2013, pp. 100–110.
5. Federal Motor Carrier Safety Administration (FMCSA). *Pocket Guide to Large Truck and Bus Statistics*, Retrieved August 11, 2015, http://www.fmcsa.dot.gov/safety/research-and-analysis/2014-pocket-guide-large-truck-and-bus-statistics.
6. NHTSA. *The Economic Burden of Traffic Crashes on Employers*. DOT HS 809 682. Retrieved August 11, 2015, http://www.nhtsa.gov/people/injury/airbags/EconomicBurden/.
7. Blower, D., and K. L. Campbell. *Large Truck Crash Causation Study (LTCCS) Analysis Series: Methodology of The Large Truck Crash Causation Study*. Federal Motor Carrier Safety Administration, FMCSA-RI-05-035. FMCSA, U.S. Department of Transportation, 2005.
8. Wickens, C. D., and J. S. McCarley. *Applied Attention Theory*. CRC press, 2007.
9. Wickens, C. D., J. Goh, J. Helleberg, W. J. Horrey, and D. A. Talleur. Attentional Models of Multitask Pilot Performance Using Advanced Display Technology. *Human Factors*, Vol. 45, No. 3, 2003, pp. 360–380.
10. Wickens, C. D., J. S. McCarley, A. L. Alexander, L. C. Thomas, M. Ambinder, and S. Zheng. Attention-Situation Awareness (A-SA) Model of Pilot Error. *Human Performance Modeling in Aviation*, 2008, pp. 213–239.
11. Horrey, W. J., C. D. Wickens, and K. P. Consalus. Modeling Drivers' Visual Attention Allocation While Interacting with In-Vehicle Technologies. *Journal of Experimental Psychology: Applied*, Vol. 12, No. 2, 2006, p. 67.
12. Fisher, D. L., and A. Pollatsek. Novice Driver Crashes: Failure to Divide Attention or Failure to Recognize Risks. In *Attention: From Theory to Practice* (Kramer, A., D. Wiegmann, and A. Kirlik, eds.), Oxford University Press, 2007, pp. 134–156.
13. Rogers, W. C., and R. R. Knipling. Commercial Driver Human Factors. *Transportation Research Circular E-C117: The Domain of Truck and Bus Safety Research*. Transportation Research Board of the National Academies, Washington, D.C., 2007.
14. Government Accountability Office. *Motor Carrier Safety. Additional Research Standards and Truck Drivers' Schedule Data could Allow More Accurate Assessments of the Hours of Service Rule*, Retrieved August 11, 2015, http://www.gao.gov/products/GAO-15-641.
15. Hanowski, R. J., J. S. Hickman, W. W. Wierwille, and A. Keisler. A Descriptive Analysis of Light Vehicle-Heavy Vehicle Interactions Using in Situ Driving Data. *Accident Analysis & Prevention*, Vol. 39, No. 1, 2007, pp. 169–179.
16. Allen, R. W., G. D. Park, and L. M. Cook. The Effect of Simulation Training on Novice Driver Accident Rates. In *Driver Behaviour and Training* (Dorn, L., ed.), Ashgate Pub. Co., Burlington, VT, 2008, pp. 265–276.

17. Kennedy, R. S., N. E. Lane, K. S. Berbaum, and M. G. Lilienthal. Simulator Sickness Questionnaire: An Enhanced Method for Quantifying Simulator Sickness. *Journal of Aviation Psychology*, Vol. 3, No. 3, 1993, pp. 203–220.
18. Pradhan, A. K., A. Pollatsek, M. Knodler, and D. L. Fisher. Can Younger Drivers be Trained to Scan for Information that will Reduce their Risk in Roadway Traffic Scenarios that are Hard to Identify as Hazardous? *Ergonomics*, Vol. 52, No. 6, 2009, pp. 657–673.
19. McKenna, F. P., and J. L. Crick. *Developments in Hazard Perception*. Transport Research Laboratory, Crowthorne, UK, 1997.
20. Underwood, G. Visual Attention and the Transition from Novice to Advanced Driver. *Ergonomics*, Vol. 50, No. 8, 2007, pp. 1235–1249.
21. Allen, R. W., G. D. Park, M. L. Cook, and D. Fiorentino. The Effect of Driving Simulator Fidelity on Training Effectiveness. *DSC 2007 North America*, Proceedings of the Driving Simulator Conference - North America, Iowa City, IA, 2007.
22. McDonald, C. C., A. H. Goodwin, A. K. Pradhan, M. R. Romoser, and A. F. Williams. A Review of Hazard Anticipation Training Programs for Young Drivers. *Journal of Adolescent Health*, Vol. 57, No. 1, 2015, pp. S15–S23.

The Standing Committee on Truck and Bus Safety (ANB70) peer-reviewed this paper (18-05168).

TRR
JOURNAL OF THE TRANSPORTATION RESEARCH BOARD

Transportation Research Record
2018, Vol. 2672(34) 39–48

DOI: 10.1177/0361198118788190
journals.sagepub.com/home/trr
SAGE

A First Investigation of Truck Drivers' Preferences and Behaviors using a Prototype Cooperative Adaptive Cruise Control System

Shiyan Yang[1,2], Steven E. Shladover[1], Xiao-Yun Lu[1], Hani Ramezani[1], Aravind Kailas[3], and Osman D. Altan[4]

Abstract

Cooperative adaptive cruise control (CACC) is a driver-assist technology that uses vehicle-to-vehicle wireless communication to realize faster braking responses in following vehicles and shorter headways compared with adaptive cruise control. This technology not only enhances road safety, but also offers fuel savings benefits as a result of reduced aerodynamic drag. The amount of fuel savings is dictated by the following distances and the driving speeds. So, the overarching goal of this work is to explore driving preferences and behaviors when following in "CACC mode," an area that remains largely unexplored. While in CACC mode, the brake and throttle actions are automated. A human factors study was conducted to investigate truck drivers' experiences and performance using CACC at shorter-than-normal vehicle following time gaps. "On-the-road" experiments were conducted by recruiting drivers from commercial fleets to operate the second and third trucks in a three-truck CACC string. The driving route spanned 160 miles on freeways in Northern California and five different time gaps between 0.6 and 1.8 seconds were tested. Factors such as cut-ins by other vehicles, road grades, and traffic conditions were found to influence the drivers' opinions about use of CACC. The findings presented in this paper provide insights into the factors that will influence driver reactions to the deployment of CACC in their truck fleets.

Cooperative adaptive cruise control (CACC) systems leverage vehicle-to-vehicle (V2V) communication based on technologies such as dedicated short range communication (DSRC) to provide coordinated longitudinal control (i.e., brake/throttle maneuvers) in vehicles, thereby enabling vehicles to automatically maintain a proper following gap behind another CACC-equipped or V2V-capable vehicle. Without V2V, CACC systems default to adaptive cruise control (ACC). Cooperative automated longitudinal control reduces delays in human response, thereby enabling shorter following distances. Increasing market penetration rate of this technology is expected to provide macro-level benefits to transportation corridors such as reducing fuel consumption and emissions (*1*), and improving traffic mobility (*2–4*).

However, the benefits of CACC in a transportation corridor will be dictated not only by the rate of adoption, but also the drivers' preferences for using the technology. For example, commercial drivers may prefer to avoid following other trucks closely, despite the technical capabilities of CACC, thereby limiting the fuel savings and throughput benefits offered by CACC. With this in mind, our work explores driver acceptance and usage of CACC during routine driving operations, especially when time gaps are much shorter than normal.

Our previous study established that drivers in passenger cars in general feel comfortable to accept a time gap less than 1 second while driving in a two-vehicle CACC string (*5*). Other studies involving ACC systems may provide insights into driver experiences when using CACC. For example, passenger car drivers in the Netherlands were favorable to using ACC on high-speed roads and in low-density traffic, but were annoyed by the occasional

[1]Partners for Advanced Transportation Technology (PATH), Institute of Transportation Studies, University of California, Berkeley, Richmond, CA
[2]Seeing Machines Ltd, Fyshwick ACT, Australia
[3]Volvo Group North America, Costa Mesa, CA
[4]Federal Highway Administration, Turner-Fairbank Highway Research Center, McLean, VA

Corresponding Author:
Address correspondence to Shiyan Yang: shiyan.yang@seeingmachines.com

clumsiness and dangerous events induced by ACC (*6*), such as hard brake because of other vehicles cutting in. The application of ACC has influenced driving behaviors such as an increased tendency to drive in the right lane (*7*) and forcing drivers to intermittently reclaim vehicle control (*8*). The work done to understand drivers' preferences and behaviors when using CACC systems is quite limited so far. Compared with passenger cars, the design and implementation of CACC systems on trucks are more involved, given that the nature of commercial fleet operation imposes rigorous requirements on drivers' concentration on the road. Although the ability of CACC to enable trucks to follow closely has been demonstrated (e.g., SARTRE, GCDC, European Truck Platooning Challenge), truck drivers' acceptance and preferences for CACC have not been investigated (*9*).

Understanding driver experiences from multiple perspectives, such as time gap preference, truck position preference, trust, and satisfaction, when operating CACC-equipped trucks on public roads in real traffic conditions is therefore extremely important. This paper presents a first investigation of driver experiences, preferences, and behaviors when driving trucks equipped with CACC. Several key aspects have been investigated:

1) Time gap preferences
2) Truck position in a CACC string
3) Using CACC with cut-ins
4) Using CACC on road grades
5) Situations during which drivers reclaim full control of the trucks

By having test drivers from commercial fleets operate CACC-equipped Class 8 Volvo tractors over 160 miles on public roads, we could document their experiences in a range of traffic conditions and road grades. While the drivers had the freedom to choose the time gap settings for automated vehicle following, the real-time truck speeds were determined by the CACC control system on the trucks. The drivers were responsible for steering, while the braking and engine control actions were automated. Although a wide range of data were collected, only the driver experience is reported in this paper. These findings provide valuable insights into the design of better CACC systems, and also aid in setting reasonable expectations for the benefits offered by CACC.

On-the-Road Experiments

Participants

Nine professional fleet truck drivers from the U.S. (7) and Canada (2) were recruited for the on-the-road experiments. All test drivers were male with an average age of 48 years. Everybody possessed a valid Class A driver license with a clean driving record and no moving violations over the past 3 years. Their driving records were verified by the UC Berkeley Fleet Services prior to commencing experiments.

The process of recruitment was hindered by the fact that the supply of truck drivers is in shortage in the U.S. (*10*). Furthermore, fleet drivers only had very limited flexibility to participate in our study because their schedules were arranged by their fleet companies according to business demands. Although a larger sample size would have been desirable, we only ended up with nine male drivers for the day time test despite 4 months of intensive recruiting efforts with assistance from our truck manufacturer partner and local trucking industry associations in California.

Trucks and CACC Control

Three Volvo Class 8 trucks (see Figure 1*a*) with an empty trailer behind each were used for the on-the-road experiment. All trucks were equipped with CACC, meaning

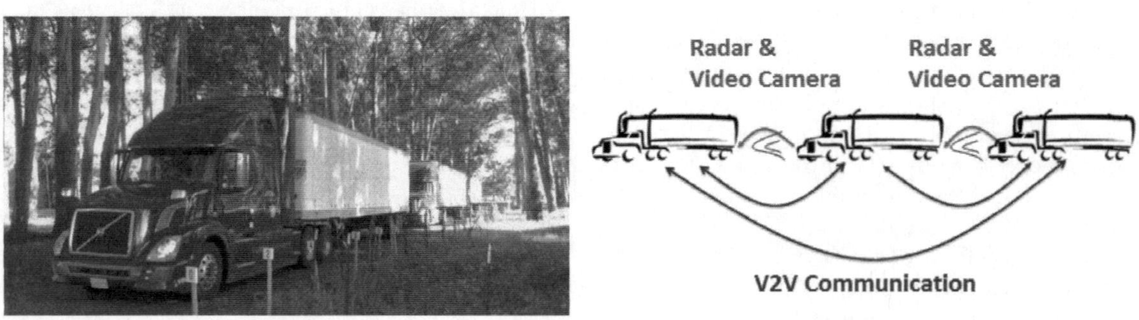

(a) (b)

Figure 1. (*a*) the Volvo Class 8 trucks and (*b*) the DSRC-based V2V communication system.

Figure 2. (*a*) the CACC control stalk and (*b*) truck cabin interior with driver–vehicle interface (DVI) and safety disengage button.

that they can exchange the control-related messages with each other via DSRC.

The control stalk behind the left side of the steering wheel, originally implemented to activate the production ACC system, was modified to activate or deactivate CACC (See Figure 2*a*). The drivers could engage or resume CACC by pushing the control button on the stalk to the left and disengage CACC by pushing it to right (see Figure 2*a*). In addition to the control stalk, the brake pedal and the red safety button (see Figure 2*b*) can be used to deactivate CACC.

Driver–Vehicle Interface (DVI)

The CACC DVI screen (see Figure 3) was redesigned in multiple iterations using QT (qt-opensource-windows-x86-msvc2015_64-5.7.1) based on field observation and human factors design guidance. Some components of the DVI, such as buttons, were edited via the online photo editor Pixlr (https://pixlr.com/editor/). The final version was implemented on a Samsung tablet that served as the DVI for the experiments. The CACC DVI presented elementary status information about other trucks in the CACC string and was used to control the time gap and driving mode (details in Table 1). The CACC interface was fitted on top of the truck instrument panel (see Figure 2*b*).

The function of each component is described in Table 1.

Time Gap

Time gaps for CACC and ACC modes are listed in Table 2. The CACC time gaps are smaller than corresponding ACC time gaps because the V2V communication function included in the CACC enables the trucks to be driven safely and smoothly at significantly shorter following time gap (or clearance distance). The CACC time gaps were chosen to match some of the time gaps that were tested for CACC on passenger cars in prior PATH research projects (*5*), which found that drivers had a significant preference for the shorter gaps. It should be noted that some of these gaps could be shorter than 30 m, which is the minimum following distance in a "caravan" as established by the California Vehicle Code. As a result, a special law was passed by the State Legislature to permit this testing to occur at distances shorter than 30 m. The lead truck only operated in the ACC mode but the two following trucks could be in CACC or ACC modes. They may switch from CACC

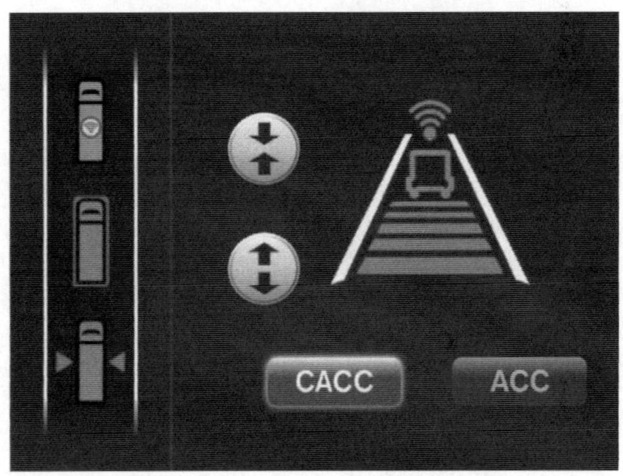

Figure 3. CACC driver–vehicle interface (DVI).

Table 1. The Function of Each Component of the DVI

Component	Description
	• The truck icons on the left side of the interface indicate the trucks that are part of a string. • The color of each truck icon body indicates the operation mode of that truck: White – Manual mode Gray – ACC mode Blue – CACC mode
	• Red icon indicates the occurrence of problems in the V2V communication system on that truck.
	• Red outline indicates that the driver has pressed the brake pedal or the foundation brakes have been applied automatically on that truck.
	• The pair of green triangles indicates the position of the host vehicle.
	• The glow indicates mode activation. In this example, CACC mode is active.
	• The five bars indicate the five CACC or ACC time gaps. Gap 1 (shortest gap)  Gap 2 (medium-to-short gap)  Gap 3 (medium gap)  Gap 4 (long-to-medium gap)  Gap 5 (longest gap) 

(continued)

Table 1. (continued)

Component	Description
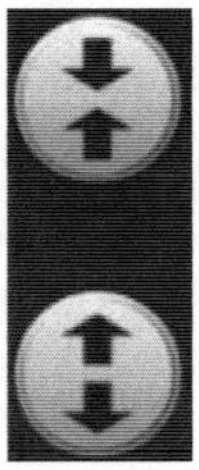	• The round buttons with arrows inside are controls to increase (bottom button) or reduce (top button) the CACC or ACC time gap. When pressing the button on the bottom, the arrows inside move away from each other; when pressing the top button, the arrows move toward each other.

Table 2. Time Gaps and Corresponding Clearance Distances in CACC and ACC Modes

Setting	CACC time gap (s)	Clearance distance (m) at 55 mph	ACC time gap (s)	Clearance distance (m) at 55 mph
1	0.6	14.8	1.1	27.0
2	0.9	22.1	1.3	32.0
3	1.2	29.5	1.5	36.9
4	1.5	36.9	1.7	41.8
5	1.8	44.3	1.9	46.7

to ACC when the V2V communication signal is not available.

Driving Tasks

The lead truck was driven by an employee from UC Berkeley with a valid Class A driver license. The test drivers drove the second and third trucks in CACC mode. They had the freedom to engage and disengage CACC and select their preferred time gaps using the DVI. However, they were responsible for steering and other maneuvers (responding to actions of other drivers) during the experiments. An experimenter sat in the front passenger seat of each following truck to monitor the CACC operations and needed to press down the safety button (see Figure 2*b*) immediately to stop CACC if it performed abnormally. But this never happened during the tests. The experimenter also needed to remind the drivers to take control of the truck in some road and traffic conditions (e.g., heavy traffic and steep downgrades).

Testing Route

The test route started from the UC Berkeley Richmond Field Station (RFS) in Richmond, via I-580 (to Emeryville), SR 24 (to Walnut Creek), I- 680 (to Pleasanton), I-580 (to Livermore), and ended around Westley on I-5 (see Figure 4). After arriving at Westley,

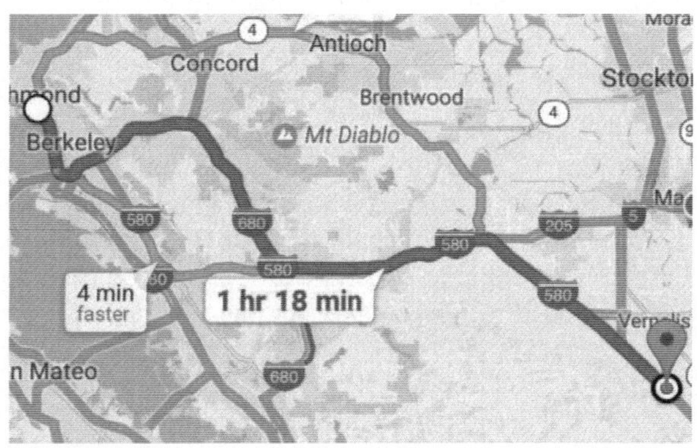

Figure 4. The testing route from RFS in Richmond to Westley.

we took a short break at a parking area near a truck stop and then returned to RFS via the same route. There is a weigh station near Livermore that drivers had to drive through when it was open.

A single trip by truck from Richmond to Westley usually takes around 1 hour 40 minutes without heavy traffic delay, so a round trip is more than 3 hours. The timing 1 hour 18 minutes on Figure 4 was calculated according to the speed of passenger cars, but the trucks were limited to the state truck speed limit of 55 mph. The on-the-road driving test normally started after 10:00 AM and ended before 2:30 PM (local time) to avoid the morning and afternoon peak congestion periods.

Experimental Procedure

First, the experimenters introduced the study to the fleet drivers using a PowerPoint presentation and video in a conference room. Also, the experimenters ensured that the drivers had enough service hours for truck driving in the study. After the introduction the drivers signed the consent form and completed the background questionnaire. Then they moved on to the training section in which they needed to get familiar with the control of CACC (e.g., engagement, disengagement, and time gap selection) and experience each time gap setting during the drive between Emeryville and Walnut Creek. Once past Walnut Creek, they had the freedom to use CACC in the way they preferred for truck platooning, but under the monitoring of the experimenter next to them. When arriving at Westley (the end of the testing route), the drivers took a short break. After the break, the driver of the second truck switched to the third truck and the driver of the third truck switched to the second truck, and then they drove back to RFS via the same route. After arriving at RFS, they finished the post-experiment debriefing questionnaire and were compensated for their time at $30 per hour, including the pre- and post-drive periods for briefings and filling out questionnaires.

Data Recording and Analysis

A PC-104 computer stored in a cabinet inside the truck cab was used to record the driver behavior data (around 100 columns) at a sampling rate of 50 Hz. However, those data are not reported in this paper because of the length limit. This paper only reports the driver experience data collected through the questionnaires. Friedman test and post hoc pairwise comparison on the time gap preference rankings were conducted using the software package R to find the most preferred time gap settings.

Results and Discussion

The results reported in this section were obtained from the final debriefing questionnaire that all the drivers filled out. The participants are designated by labels P1 to P9 in which the letter "P" represents "Participant." The drivers' written answers in the tables in this section were processed by the experimenter to ensure their grammatical correctness and readability. In the questionnaires, a 7-point scale was used to quantify drivers' subjective reactions (e.g., familiarity, comfort, trust, and reliability) to each item, with 1 representing "not familiar/comfortable/trustable/reliable at all" and 7 representing "very familiar/comfortable/trustable/reliable", and with the other 5 levels evenly distributed between the two extremes.

Drivers' Demographics

The demographics of the drivers was collected by the background questionnaire and reported in Table 3.

On average, the tested drivers are relatively senior in the trucking business, with an average of 21.1 years of driving experience; most were for a company or fleet versus driving as an owner or operator. The drivers have more experience in manual transmission driving than automatic transmission driving and they drive more in short haul compared with long haul. Their driving mileages are largely on freeway and partially on highway or urban. However, they only have very limited experience with ACC, collision warning systems, and driving in truck platoons, in their relatively long careers as commercial vehicle operators.

Table 3. Demographics of the Recruited Drivers

Background question	Mean (Std. dev.)
Age	48 (13)
Gender	9 male, 0 female
How many years have you driven tractor-trailer trucks?	21.1 (14.1)
How many years of experience do you have as an owner/operator?	2.4 (3.7)
How many years of experience do you have as a company or fleet driver?	18.9 (11.9)
What fraction of your heavy truck driving is with manual versus automatic transmission?	75.9 (31.0) Manual 24.1 (31.0) Automatic
What fraction of your driving is short haul versus long haul?	66.1 (34.4) Short haul 33.9 (34.4) Long haul
What is the fraction of total mileage you spend driving on freeways, other highways, and urban streets in a typical month?	63.3 (20.0) Freeway 15.0 (14.6) Highway 21.7 (17.7) Urban
How is your familiarity with Adaptive Cruise Control (ACC)?	1.4 (1.9)/7
How is your familiarity with collision warning systems?	2.1 (2.0)/7
How is your familiarity with driving in a truck platoon?	0.7 (2.0)/7

Table 4. Driving Experiences in Different Positions in the String

Question[a]	Results
Did you notice a difference between driving in the 2nd or 3rd position?	• 5 No • 3 Yes
If you notice a difference between driving in the 2nd and 3rd position, what's the difference?	• The 3rd truck provided clearer view of the highway to anticipate events (P1) • The 3rd truck's brake control was not as good as the 2nd truck's (P3, P4)
Did you have a preference for being in the 2nd or 3rd position?	• 2 preferred 2nd position • 1 preferred 3rd position • 5 no preference
If you prefer 2nd position or 3rd position, please explain why you have such a preference.	• Preferred 3rd position because it enabled better vantage point and more anticipation of the on-road events because of the view of both other trucks (P1) • Preferred the 2nd truck because of its better control reaction to the lead truck (P3, P4)

[a]Participant 9 was not able to answer these questions because he only drove the second truck in a two-truck string without experiencing the third truck.

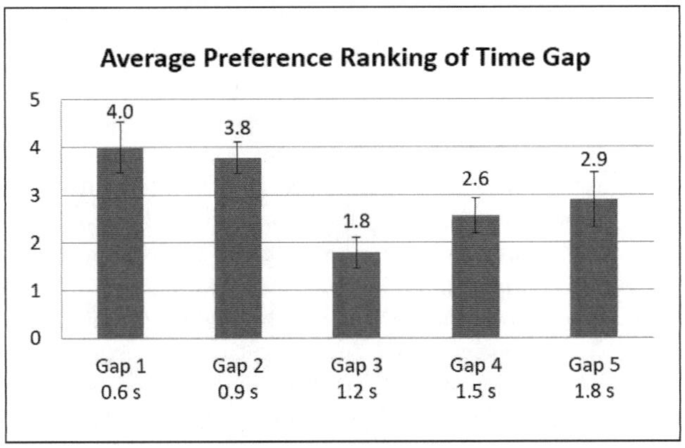

Figure 5. Average preference rankings of time gap at each setting. Smaller value indicates higher preference ranking.

Time Gap Preference Ranking

The drivers ranked the five time gaps from 1 to 5 to indicate their preference for each time gap from high to low. The Friedman test showed that the average preference rankings of the time gaps were significantly different from each other ($\chi^2 = 10.1$, df = 4, $p = 0.040$; see Figure 5). Post hoc test showed that the preference ranking of Gap 3 (1.8) was significantly higher than that of Gap 1 (4.0, $p < 0.001$), Gap 2 (3.8, $p < 0.001$), and Gap 5 (2.9, $p = 0.027$), but not significantly different from Gap 4 (2.6). Similarly, the preference ranking of Gap 4 was significantly higher than that of Gap 1 ($p = 0.016$) and Gap 2 ($p = 0.045$).

Participants preferred the Gap 3 (1.2 s) and Gap 4 (1.5 s) settings of CACC the most. When the time gap was very small, their view of the road ahead was obstructed by the trailer of the preceding truck. This was reflected in the debriefing questionnaire, where eight out of nine drivers chose the two shortest CACC time gaps (0.6 s and 0.9 s) as the gaps where they felt it was most difficult to see enough of the road ahead for comfortable driving. The drivers also did not prefer Gap 5, the largest CACC time gap. They mentioned that the large gap tended to encourage cut-ins by other vehicles on the road, mainly light duty vehicles. Therefore, the two medium gap settings (Gap 3 and Gap 4) were preferred by the drivers as a compromise between their perceived driving safety/comfort and their intention to deter cut-ins by other vehicles.

Truck Position in the CACC String

Drivers' preference for their position in the string is described in Table 4. The position of a truck in the CACC string was not a key factor limiting or enhancing drivers' ability to see ahead. Only one participant reported that driving in the third truck provided a better view of the highway and more anticipation of the events ahead. He thus preferred driving the last truck. Two other drivers noticed the difference in the braking performance rather than the road visibility between the following trucks. They found that the second truck had better braking performance than the third truck so that they preferred the second truck. The majority of the drivers (5 out of 8) didn't notice any difference between driving in the second and third trucks and had no preference for either position.

CACC Response to Cut-Ins and Road Grade

During the experiments, the CACC system responded to a vehicle cut-in by detecting it and then slowing the truck to leave a larger distance between itself and the cut-in vehicle. Once the vehicle cuts out, the truck increased its speed automatically to reduce its distance to the preceding truck

Table 5. Experience of Using CACC in Different Driving Conditions

Category	Question	Results mean (Std. dev.)
Cut-in	When a vehicle cut in between you and the truck ahead of you, how comfortable did you feel with the CACC system response?	5.2 (2.1)/7
	How much did you trust the CACC system to ensure safety when a cut-in occurred ahead of you?	5.0 (1.8)/7
Road grade	How reliably do you think the CACC worked when you drove on upgrades?	4.6 (2.0)/7
	How reliably do you think the CACC worked when you drove on downgrades?	3.1 (1.8)/7

until it reached the time gap (or corresponding following distance) set by CACC. When driving uphill or downhill the CACC system needed to manipulate engine and braking controls to offset the acceleration and deceleration caused by the road grades and maintain the proper following time gap. The drivers' opinions about the CACC response to cut-ins and road grade are reported in Table 5.

On average the participants felt comfortable with CACC response to cut-in vehicles. Also, they seemed to be convinced of the safety benefits offered by CACC, especially during vehicle cut-ins. However, participants were less confident with the reliability of the prototype CACC when it was operating on steep road grades, especially on downgrades. This was partially attributable to the pre-test instructions and partially attributable to their direct experience. In particular, the prototype CACC control relies primarily on engine braking for deceleration, which cannot generate sufficient deceleration to slow the truck on steeper downgrades. When the foundation brakes were activated to provide stronger deceleration, that disengaged the CACC control based on some of the internal logic of the Volvo ACC system that could not be circumvented. On upgrades, CACC sometimes could not trigger enough acceleration for the following truck to stay close to the preceding trucks. During the pre-test briefing, experimenters explained these limitations to the participants and asked them to disengage CACC whenever they were not comfortable with its performance, including on steep grades.

Switching from Automation to Manual Driving

During the experiments, the drivers were allowed to turn off the CACC system and take complete control of the truck as necessary. Eight of the drivers reported deactivating the CACC system because of heavy (high density) traffic, which was by far the most significant cause of driver interventions. Heavy traffic increases the likelihood of vehicle cut-ins, and frequent cut-ins can disrupt driving in CACC strings. Furthermore, the trucks may be slowed down by heavy traffic, which results in shorter following distances based on the unchanged time gap setting, which could be uncomfortable for the drivers. On the other hand, the drivers demonstrated greater willingness to use CACC when traffic was light and predictable, such as driving on the rural highway I-5.

Road grade was the second most frequently mentioned factor that contributed to the transition to manual driving (mentioned by four drivers) because of the performance limitations of the current CACC prototype implementation that was not able to provide sufficiently strong and smooth control to counteract the accelerations and decelerations caused by the road grades.

Highway merging was also a situation in which drivers preferred operating in manual mode, and one driver reported deactivating the CACC for a highway merge. The automated truck string may "block" other vehicles when they had to merge into the highway after passing the limited length of on-ramp. Since the sensors in the current CACC prototype were not able to detect the merging vehicles approaching alongside the trucks, drivers had to switch to the manual mode to let them cut in, or change lane to provide enough space for other vehicles to merge into highway, or speed up to pass the merging vehicles.

Limitations of Prototype CACC System

Keeping in mind that that CACC implementation was an advanced research prototype, there were limitations of the performance of the CACC system on the trucks. From driver experience of operating the system on public highways, a variety of concerns were reported. They are unreliability and jerkiness in the speed control (P2, P7), wireless communication errors (P3), CACC reliability concerns (P5), limited road visibility from the following truck (P7), and the position of the tablet display being outside peripheral vision (P8). For example, the brake control by CACC using engine braking was not strong enough, so that the second truck could get too close to the lead truck on downhill sections (P2). Sometimes the truck failed to release brakes in time so that its following distance to the preceding truck became much larger than what it should be (P2). Furthermore, they mentioned a few potential issues for CACC that should be explored in the future, including the impact of mechanical breakdown or tire blowout on the preceding truck to the following truck (P1), highway infrastructure to assist truck platoon (P8), and CACC-induced complacency (P4).

Overall Experience with CACC

On the 7-point scale used in the post-experiment questionnaire, the drivers reported their overall satisfaction with the CACC system (mean value of 5.6, with a standard deviation of 0.9). These results show that the drivers were satisfied with their truck driving experience with the driver-assist capabilities of CACC. However, CACC failed to make the commercial vehicle operation job more attractive to them (rating for increasing attractiveness of truck driving job at a mean value of only 0.6 on the 7-point scale), which implies that other factors, beyond automation technology, will have to change to increase job satisfaction. They also reported that it did not change their attention to the driving task in any significant way (mean value of 0.1 on the 7-point scale).

Conclusion

This paper presented a first human factors study on truck drivers' on-the-road experience of using CACC. Although the sample size was not ideally large, the findings of this study still illustrated important issues for drivers when they interacted with CACC and demonstrated the diversity of driver preference for CACC. A large sample size would provide higher statistical confidence in the results but may not be likely to change the findings in fundamental ways.

Our findings provide important insights into the acceptance of the technology at a higher level, as well as revealing preferences with respect to its usage (for example, gap settings, usage on road grades, etc.). It was evident that the test drivers did not prefer to drive too close (< 0.9 s) or far (1.8 s) behind the lead trucks, and seemed to prefer time gaps of 1.2 and 1.5 seconds. While the shorter time gaps limited their driving view from following too closely, the largest time gap seemed to encourage more frequent vehicle cut-ins.

The test drivers did not have a preference regarding the position of the following truck (second or third in the three-truck string), which should allow for flexibility in forming ad-hoc CACC strings "on the fly". Overall, the drivers felt comfortable with the CACC system, but preferred the manual mode in cases of heavy traffic and merging on the highway. Despite the performance limitations of the prototype CACC system, the positive feedback from the drivers based on their experience driving on public roads in different traffic and road conditions inspires confidence for such advanced technologies in the trucking industry.

Acknowledgments

This research was supported by the Federal Highway Administration's Exploratory Advanced Research Program under Cooperative Agreement No. DTFH61-13-R-00011, with cost sharing by the State of California Transportation Agency, Department of Transportation (Caltrans). We would like to thank the California Trucking Association for its assistance with driver recruiting. We would also like to thank our PATH colleagues John Spring and David Nelson for providing important technical support for the on-the-road experiment.

Author Contributions

The authors confirm contribution to the paper as follows: study conception and design: Shiyan Yang, Steven Shladover, Xiao-Yun Lu; data collection: Shiyan Yang, Xiao-Yun Lu, Aravind Kailas, Hani Ramezani; analysis and interpretation of results: Shiyan Yang, Steven Shladover; draft manuscript preparation: Shiyan Yang, Steven Shladover, Aravind Kailas, Osman Altan. All authors reviewed the results and approved the final version of the manuscript.

References

1. Browand, F., J. McArthur, and C. Radovich. *Fuel Saving Achieved in the Field Test of Two Tandem Trucks*. California PATH Research Report, Berkeley, CA, 2004.
2. Arnaout, G. M., and J. P. Arnaout. Exploring the Effects of Cooperative Adaptive Cruise Control on Highway Traffic Flow using Microscopic Traffic Simulation. *Transportation Planning and Technology*, Vol. 37, No. 2, 2014, pp. 186–199. https://doi.org/10.1080/03081060.2013.870791.
3. Lunge, A., and P. Borkar. A Review on Improving Traffic Flow using Cooperative Adaptive Cruise Control System. *Proc., 2015 2nd International Conference on Electronics and Communication Systems (ICECS)*, Coimbatore, India, IEEE, New York, NY, pp. 1474–1479.
4. Shladover, S., D. Su, and X. Y. Lu. Impacts of Cooperative Adaptive Cruise Control on Freeway Traffic Flow. *Transportation Research Record: Journal of the Transportation Research Board*, 2012. 2324: 63–70.
5. Nowakowski, C., J. O'Connell, S. E. Shladover, and D. Cody. Cooperative Adaptive Cruise Control: Driver Acceptance of Following Gap Settings Less than One Second. *Proceedings of the Human Factors and Ergonomics Society Annual Meeting*, Vol. 54, No. 24, 2010, pp. 2033–2037.
6. de Winter, J. C. F., C. M. Gorter, W. J. Schakel, and B. van Arem. Pleasure in using Adaptive Cruise Control: A Questionnaire Study in the Netherlands. *Traffic Injury Prevention*, Vol. 18, No. 2, 2017, pp. 216–224. https://doi.org/10.1080/15389588.2016.1220001.

7. Strand, N., J. Nilsson, I. C. M. Karlsson, and L. Nilsson. Exploring End-User Experiences: Self-Perceived Notions on Use of Adaptive Cruise Control Systems. *IET Intelligent Transport Systems*, Vol. 5, No. 2, 2011, pp. 134–140.
8. Larsson, V., L. Johannesson, B. Egardt, and A. Lassson. Benefit of Route Recognition in Energy Management of Plug-In Hybrid Electric Vehicles. *Proc., 2012 American Control Conference (ACC)*, Montreal, Quebec, Canada, IEEE, New York, NY, 2012, pp. 1314–1320.
9. Bergenhem, C., S. Shladover, E. Coelingh, C. Englund, and S. Tsugawa. Overview of Platooning Systems. *Presented at 19th ITS World Congress*, Vienna, Austria, 2012.
10. Costello, B., and R. Suarez. *Truck Driver Shortage Analysis 2015*. The American Trucking Associations, Arlington, VA, 2015.

The Standing Committee on Truck and Bus Safety (ANB70) peer-reviewed this paper (18-01305).

Article

TRR
JOURNAL OF THE TRANSPORTATION RESEARCH BOARD

Transportation Research Record
2018, Vol. 2672(34) 49–62

DOI: 10.1177/0361198118776815
journals.sagepub.com/home/trr
SAGE

A Comparative Analysis of Factors Affecting the Frequency and Severity of Freight-Involved and Non-Freight Crashes on a Major Freight Corridor Freeway

Samuel G. Taylor[1], Brendan J. Russo[1], and Emmanuel James[1]

Abstract
Traffic crashes cost society billions of dollars each year as a result of property damage, injuries, and fatalities. Additionally, traffic crashes have a negative impact on mobility, as they are a primary cause of non-recurring delay. With the Interstate 10 corridor between the ports of Los Angeles and Houston being one of the most vital links for goods movement across the United States, safety and mobility along this freeway, particularly for freight traffic, are of significant concern. This study, which utilized six years of crash data from the state of Arizona, explores factors affecting the frequency and severity of crashes along the Arizona portion of the I-10 corridor, with a particular focus on freight-related crashes. The safety performance along the I-10 is analyzed through the development of crash frequency and severity prediction models using integrated crash, roadway, traffic, and environmental data. Negative binomial and ordered logit models, with the incorporation of random parameters, were estimated to provide a detailed understanding of factors associated with freight-involved crashes and how they compare to non-freight crashes in terms of frequency and severity. The results showed that several roadway- crash-, vehicle-, and person-related variables were associated with the frequency and/or severity of crashes along the study corridor. These findings provide important insights which can be used to develop or plan countermeasures aimed at improving the safety and efficiency of freight travel, which may include new ITS technologies, and targeted educational and enforcement campaigns.

Goods movement across the US is one of the most significant factors for economic growth in the United States. Between 1993 and 2002, the national gross domestic product (GDP) increased by 33% while the value of freight shipments increased by 45% (*1*). In 2013, the US transportation system moved a daily average of about 55 million tons of freight, valued at more than $49.3 billion, with trucks transporting about 70% of that total (*2*).

The country has reached an important crossroads: with more people and goods taking to the roads, motor vehicle deaths on the nation's roadways are on a historic 14% rise from 2014 to 2016 (*3*). Additionally, national crash statistics from 2015 show that a larger percentage of large truck and bus crashes result in fatalities than other crash types (*4*). The National Safety Council estimates the current comprehensive cost of a motor vehicle death to be $10,082,000 (*5*). Given these recent statistics, it is clear there is a strong need for improving traffic safety measures on vital freight corridors.

Many states have already begun to take steps towards reducing friction between passenger and commercial vehicles. For example, the I-95 Corridor Coalition involves 15 different states that all work together to share valuable information with the shared goal of enhancing mobility, safety, and efficiency between each state (*6*). In Arizona, the newly formed I-10 Corridor Coalition shares a similar goal with California, New Mexico, and Texas (*7*). Both the west and east sections of the I-10 in Arizona are considered by the Arizona Department of Transportation (ADOT) to be vital for the overall health of the statewide transportation system (*8*). However, no in-depth analysis of freight-involved

[1]Department of Civil Engineering, Construction Management, and Environmental Engineering, Northern Arizona University, Flagstaff, AZ

Corresponding Author:
Address correspondence to Brendan J. Russo: brendan.russo@nau.edu

crashes on the entirety of Arizona's portion of the I-10 has been completed. This study aims to fill that void and provide important information which may be useful in planning future traffic safety improvements.

Several studies have been completed across the US and Canada that have examined factors affecting either the frequency or severity of large truck crashes with most focusing primarily on injury severity as a function of crash reported variables. In 2015, a study was completed in Ontario, Canada that used a general estimating equation model to compare frequency predictions for truck-involved and non-truck-involved crashes (*9*). The study concluded that wider lane widths increased frequency, higher truck percentage decreased frequency, and higher speed limits also decreased frequency in truck-involved crashes. Additionally, it was found that in two successive years, at any given location, there was no direct correlation between truck-involved crashes and non-truck-involved crashes (*9*).

Another study, which was completed in 2017, used multinomial logit and negative binomial models under the Bayesian estimation framework to analyze crash severity and frequency, respectively (*10*). They found that inclement weather conditions increased the frequency and severity of truck-involved crashes (*10*). The results also showed that higher speed limits reduced the frequency of truck crashes and that dark lighting conditions and rural terrains increased the severity of truck-involved crashes (*10*). Significant safety needs were highlighted in two 2017 I-10 performance reports by ADOT, in which 11 out of 24 study segments were identified as having a "high" need for safety improvements (*8*, *11*).

To gain further insights, this study utilizes crash data from 2010 to 2015, roadway and traffic characteristics, and two different statistical approaches to analyze factors which may be associated with high crash frequencies and severe injuries on the I-10 in Arizona. With the entire freeway through Arizona being combined and evaluated as a freight corridor, the findings of this study may be used for future safety improvements to enhance freight safety and mobility.

Data Description

The data for this study were acquired from ADOT and included a record of all reported crashes in the state of Arizona from 2010 to 2015, as well as geometric and traffic volume data from ADOT's Multimodal Planning Division (MPD). These data were filtered to only include incidents and roadway characteristics on the I-10 through Arizona. To gain a full understanding of crashes on the I-10 and to satisfy the data needs for the frequency models and severity models, three different datasets were created.

The first dataset was designed for the frequency model and consisted of 264 predefined segments with separate segments for the eastbound and westbound directions on the I-10. Annual crashes were then linked to the segments by GPS coordinates provided in the crash reports, resulting in a total of 1,584 segment-year cases. The crashes assigned to each segment were classified as "freight-involved" or "non-freight" by filtering crashes with five unique body-style identifiers that encompass tractor-trailers, box trucks, and auto carriers.

Geometric characteristics were then assigned to each segment by intersecting several different geometric layers in ArcMap (*12*) with the predefined study segments. Because the purpose of this study was to compare factors affecting freight-involved crashes to factors affecting non-freight crashes, a separate dependent variable was incorporated into the dataset that included all non-freight-involved crashes on the I-10 over the study period. These crash counts were also linked to the same segments by using ArcMap. For this study, it was assumed that the geometric characteristics for each segment stayed constant during the study period. However, the Annual Average Daily Traffic (AADT) counts and truck percent varied from year to year. Ultimately, the 'freight-involved' dataset included 5,695 crashes while the 'non-freight' dataset (which included all non-freight crashes on the I-10 over the study period) included 30,037 crashes.

During the frequency modeling process defined in the following section, many of the continuous geometric variables such as number of lanes, median width, and median type were reclassified as binary indicator variables (0 or 1). Since the study section of the I-10 spans from the western border to the eastern border of Arizona and passes through urban areas with high traffic volumes and unique geometric and road user characteristics, it is important to account for this variation in the model. This was accomplished by identifying each segment in the urban Phoenix area with a high occupancy vehicle lane as the indicator variable "Phoenix." Descriptive statistics for variables utilized in the frequency models are presented in Table 1.

For the datasets utilized in the severity analyses, each person in each crash on the I-10 was recorded as a unique observation. Geometric characteristics were again assigned using ArcMap and the GPS coordinates of the crash that each individual person was involved in. The freight-involved severity dataset used the same freight classifications described previously and consisted only of persons in a freight-involved crash (a total of 14,148 observations). The non-freight severity dataset consisted of all persons involved in crashes on the I-10, excluding those that were freight-involved, for a total of 71,051 observations. For the severity models, the ordered

Table 1. Descriptive Statistics for Frequency Model Variables

Roadway characteristics	Mean	Std. dev.	Min.	Max.
Segment length (miles)	2.96	3.05	0.19	15.68
AADT	45,848	38,963	4,350	171,154
Ln(AADT)	10.32	0.95	8.38	12.05
Cable barrier*	0.10	0.30	0.00	1.00
Concrete barrier*	0.30	0.46	0.00	1.00
No barrier*	0.61	0.49	0.00	1.00
Median width <39 Ft*	0.27	0.44	0.00	1.00
Median width 40–79 Ft*	0.58	0.49	0.00	1.00
Median width >80 Ft*	0.16	0.37	0.00	1.00
Right shoulder (Ft)	10.64	1.93	6.63	22.71
Left shoulder (Ft)	7.63	3.70	3.00	18.12
3 or 4 Lanes*	0.41	0.49	0.00	1.00
5 or 6 Lanes*	0.16	0.37	0.00	1.00
2 Lanes*	0.42	0.49	0.00	1.00
Speed limit 45 or 55*	0.12	0.32	0.00	1.00
Speed limit 65*	0.42	0.49	0.00	1.00
Speed limit 75*	0.46	0.50	0.00	1.00
Degree of curvature	0.02	0.17	0.00	1.68
Percent grade	0.74	0.53	0.00	2.22
Truck percent	13.70	11.10	1.82	54.66
Phoenix indicator*	0.30	0.46	0.00	1.00
Freight-involved crashes[a]	3.60	3.65	0.00	24.0
Non-Freight crashes[b]	19.60	26.53	0.00	243

*Binary indicator variable (i.e., 0 or 1).
[a]Dependent variable for the freight-involved crashes only model.
[b]Dependent variable for the non-freight model.

discrete variable, injury status, was modeled as the dependent variable. The injury status levels are described by the Arizona Crash Report Forms Instruction Manual as follows (*13*):

- 5-Injury: Fatal Injury (K-injury)
- 4-Injury: Incapacitating Injury (A-injury)
- 3-Injury: Non-Incapacitating Injury (B-injury)
- 2-Injury: Possible Injury (C-injury)
- 1-No Injury (O—No injury)

In the two severity datasets, all variables were recoded into binary indicator variables. For example, the "summer months" variable takes the value of 1 if the crash occurred during a summer month (i.e., June–August), otherwise, it takes the value of 0. Also of note, if a variable was initially recorded as "unknown" or "not reported" then the entire observation was omitted from the final models. Unlike the frequency models, the severity model framework allows for incident-, unit-, and person-level characteristics, as well as the geometric characteristics to be included as independent variables. The body style indicator variables for persons in freight vehicles and persons in passenger vehicles are important in this study as they provide insight into the safety impact of large trucks sharing the road with smaller passenger vehicles. Summary statistics for the freight-involved severity dataset are presented in Table 2 and summary statistics for the non-freight severity dataset are presented in Table 3. Again, note that all variables contained in Tables 2 and 3 are binary indicator variables.

Analysis Methodology

Two different types of model were used for this study. The first was a random parameters (RP) negative binomial regression model, used for analyzing factors which may affect the frequency of crashes. The second was an RP ordered logit model, used for analyzing factors which may affect the severity of crashes (in terms of injury outcomes). The two-model approach and the combination of different data sources helps create a detailed understanding of factors involved in freight crashes and non-freight crashes on a major freight corridor freeway.

Crash Frequency Model: Negative Binomial Regression

Using a negative binomial model for crash frequency predictions has been proven to be effective in past studies (*10*, *14*, *15*), and it is one of the most popular methods for the development of Safety Performance Functions

Table 2. Summary Statistics for Freight-Involved Severity Model Variables

	Freight-involved crash occupant injuries by severity level					
*Total observations = 14,148	No injury	C-injury	B-injury	A-injury	K-fatal	Total
Environmental characteristics						
Summer months	2,877 (83%)	232 (7%)	259 (8%)	42 (1%)	36 (1%)	3,446
Other months	8,966 (83%)	735 (7%)	797 (7%)	134 (1%)	70 (<1%)	10,702
Blowing sand and/or dust	140 (65%)	16 (7%)	38 (18%)	17 (8%)	6 (3%)	217
Other weather conditions	11,652 (84%)	948 (7%)	1,014 (7%)	159 (1%)	100 (<1%)	13,873
Dark light conditions	3,015 (80%)	295 (8%)	364 (10%)	64 (2%)	44 (1%)	3,782
First harmful event						
Collision with concrete barrier	89 (73%)	10 (8%)	17 (14%)	4 (3%)	2 (2%)	122
Rollover	146 (43%)	38 (11%)	126 (37%)	18 (5%)	12 (4%)	340
Jackknife	20 (91%)	2 (9%)	0 (0%)	0 (0%)	0 (0%)	22
Other first harmful events	11,586 (85%)	917 (7%)	913 (7%)	154 (1%)	92 (<1%)	13,662
Collision manner						
Single vehicle	826 (77%)	54 (5%)	160 (15%)	19 (2%)	8 (1%)	1,067
Angle	380 (75%)	57 (11%)	53 (10%)	12 (2%)	5 (1%)	507
Head on	49 (53%)	13 (14%)	7 (8%)	9 (10%)	15 (16%)	93
Sideswipe same direction	4,577 (91%)	223 (4%)	210 (4%)	23 (<1%)	13 (<1%)	5,046
Sideswipe opposite direction	25 (69%)	1 (3%)	8 (22%)	1 (3%)	1 (3%)	36
Other collision manners	4,549 (79%)	563 (10%)	543 (9%)	78 (1%)	37 (<1%)	5,770
Body style						
Freight vehicle	6,404 (89%)	282 (4%)	397 (6%)	57 (1%)	33 (<1%)	7,173
Passenger vehicle	4,966 (78%)	631 (10%)	600 (9%)	99 (2%)	62 (1%)	6,358
Motorcycle	7 (23%)	2 (7%)	10 (33%)	8 (27%)	3 (10%)	30
Other vehicle	206 (89%)	14 (6%)	8 (3%)	4 (2%)	0 (0%)	232
Event sequence						
Cross median	35 (41%)	10 (12%)	24 (28%)	7 (8%)	10 (12%)	86
Run-off-road right	505 (60%)	82 (10%)	192 (23%)	31 (4%)	30 (4%)	840
Run-off-road left	567 (57%)	129 (13%)	233 (23%)	42 (4%)	21 (2%)	992
Other event sequences	10,807 (87%)	762 (6%)	646 (5%)	100 (1%)	47 (<1%)	12,362
Age and gender						
Age 24 or less	2,420 (82%)	219 (7%)	254 (9%)	37 (1%)	29 (1%)	2,959
Age 65 or up	730 (79%)	65 (7%)	86 (9%)	25 (3%)	16 (2%)	922
Other ages	8,458 (85%)	666 (7%)	706 (7%)	113 (1%)	61 (<1%)	10,004
Female	3,351 (78%)	465 (11%)	389 (9%)	72 (2%)	28 (1%)	4,305
Other genders	8,368 (86%)	501 (5%)	666 (7%)	104 (1%)	78 (<1%)	9,717
Safety device and violation						
Safety device used	11,106 (85%)	895 (7%)	886 (7%)	122 (1%)	43 (<1%)	13,052
Drugs or alcohol used	56 (41%)	8 (6%)	18 (13%)	11 (8%)	42 (31%)	135
Roadway characteristics						
Median width <20 Ft	2,357 (87%)	191 (7%)	125 (5%)	15 (1%)	6 (<1%)	2,694
Median width >80 Ft	1,548 (78%)	132 (7%)	238 (12%)	40 (2%)	35 (2%)	1,993
Other median widths	7,938 (84%)	644 (7%)	693 (7%)	121 (1%)	65 (<1%)	9,461
Speed limit 75	3,563 (79%)	241 (5%)	510 (11%)	116 (3%)	75 (2%)	4,505
Other speed limits	8,280 (86%)	726 (7%)	546 (6%)	60 (<1%)	31 (<1%)	9,643
Right shoulder width <10 Ft	1,668 (78%)	129 (6%)	267 (12%)	51 (2%)	28 (1%)	2,143
Other right shoulder widths	10,175 (85%)	838 (7%)	789 (7%)	125 (1%)	78 (<1%)	12,005
Left shoulder width <4 Ft	3,176 (80%)	230 (6%)	406 (10%)	76 (2%)	60 (2%)	3,948
Other left shoulder widths	8,667 (85%)	737 (7%)	650 (6%)	100 (1%)	46 (<1%)	10,200
Percent of trucks >20%	1,012 (75%)	73 (5%)	195 (14%)	37 (3%)	28 (2%)	1,345
Level roadway	11,039 (84%)	905 (7%)	950 (7%)	154 (1%)	94 (1%)	13,142
Other roadway grade	767 (80%)	56 (6%)	103 (11%)	22 (2%)	12 (1%)	960
Grand total	11,843 (84%)	967 (7%)	1056 (7%)	176 (1%)	106 (<1%)	14,148

(i.e., crash prediction models) with rare events (*16*). The negative binomial regression model is derived from the general form of the Poisson regression model, with the Poisson parameter being rewritten as shown in Equation 1 (*17*):

$$\lambda_i = EXP(\beta_0 + \beta_1 X_1 + \beta_i X_i + \varepsilon_i) \quad (1)$$

Where:

λ_i: Poisson parameter for road segment i (i.e., predicted number of annual crashes for road segment i)

Table 3. Summary Statistics for Non-Freight Severity Model Variables

	Non-freight crash occupant injuries by severity level					
*Total observations = 71,051	No injury	C-injury	B-injury	A-injury	K-fatal	Total
Environmental characteristics						
Summer months	13,356 (83%)	1,308 (8%)	1,230 (8%)	225 (1%)	69 (<1%)	16,188
Other months	45,563 (83%)	4,689 (9%)	3,816 (7%)	634 (1%)	161 (<1%)	54,863
Blowing sand and/or dust	179 (82%)	7 (3%)	32 (15%)	0 (0%)	0 (0%)	218
Other weather conditions	58,605 (83%)	5,975 (8%)	5,000 (7%)	855 (1%)	229 (<1%)	70,664
Dark light conditions	13,476 (80%)	1,416 (8%)	1,514 (9%)	334 (2%)	116 (<1%)	16,856
First harmful event						
Collision with concrete barrier	1187 (67%)	251 (14%)	294 (16%)	44 (2%)	8 (<1%)	1,784
Rollover	946 (35%)	399 (15%)	975 (36%)	285 (10%)	112 (4%)	2,717
Jackknife	108 (94%)	4 (3%)	3 (3%)	0 (0%)	0 (0%)	115
Other first harmful events	56,676 (85%)	5,343 (8%)	3,774 (6%)	530 (<1%)	110 (<1%)	66,433
Collision manner						
Single vehicle	8,587 (72%)	953 (8%)	1,812 (15%)	457 (4%)	148 (1%)	11,957
Angle	1,299 (80%)	156 (9%)	162 (10%)	30 (2%)	0 (0%)	1,647
Head on	198 (62%)	40 (13%)	50 (16%)	16 (5%)	15 (5%)	319
Sideswipe same direction	10,257 (90%)	586 (5%)	436 (4%)	54 (<1%)	12 (<1%)	11,345
Sideswipe opposite direction	102 (80%)	12 (9%)	11 (9%)	2 (2%)	0 (0%)	127
Other collision manners	35,902 (84%)	4,117 (10%)	2,406 (6%)	257 (<1%)	28 (<1%)	42,710
Body style						
Passenger vehicle	53,916 (83%)	5,551 (8%)	4,468 (7%)	674 (1%)	191 (<1%)	64,800
Motorcycle	118 (19%)	84 (14%)	288 (47%)	104 (17%)	15 (2%)	609
Other vehicle	1,616 (91%)	82 (5%)	49 (3%)	24 (1%)	11 (<1%)	1,782
Event sequence						
Cross median	215 (56%)	49 (13%)	74 (19%)	29 (8%)	15 (4%)	382
Run-off-road right	2,578 (59%)	488 (11%)	926 (21%)	265 (6%)	90 (2%)	4,347
Run-off-road left	3,128 (61%)	573 (11%)	1,044 (20%)	267 (5%)	119 (2%)	5,131
Other event sequences	53,436 (86%)	5,004 (8%)	3,245 (5%)	377 (<1%)	43 (<1%)	62,105
Age and gender						
Age 24 or less	19,227 (84%)	1,641 (7%)	1,622 (7%)	259 (1%)	52 (<1%)	22,801
Age 65 or up	3,207 (81%)	307 (8%)	327 (8%)	62 (2%)	46 (1%)	3,949
Other ages	35,444 (82%)	3,979 (9%)	3,067 (7%)	536 (1%)	132 (<1%)	43,158
Female	25,202 (80%)	3,287 (10%)	2,420 (8%)	375 (1%)	92 (<1%)	31,376
Other genders	33,374 (85%)	2,706 (7%)	2,624 (7%)	484 (1%)	138 (<1%)	39,326
Safety device and violation						
Safety device used	56,811 (84%)	5,666 (8%)	4,454 (7%)	542 (<1%)	79 (<1%)	67,552
Drugs or alcohol used	385 (52%)	60 (8%)	134 (18%)	46 (6%)	115 (16%)	740
Roadway characteristics						
Median width <20 Ft	13,416 (85%)	1,335 (8%)	925 (6%)	123 (<1%)	26 (<1%)	15,825
Median width >80 Ft	3,742 (75%)	414 (8%)	635 (12%)	181 (4%)	48 (1%)	5,020
Other median widths	41,761 (83%)	4,248 (8%)	3,486 (7%)	555 (1%)	156 (<1%)	50,206
Speed limit 75	9,483 (78%)	792 (6%)	1,405 (12%)	374 (3%)	139 (1%)	12,193
Other speed limits	49,436 (84%)	5,205 (8%)	3,641 (6%)	485 (<1%)	91 (<1%)	58,858
Right shoulder width <10 Ft	4,895 (76%)	502 (8%)	760 (12%)	203 (3%)	67 (1%)	6,427
Other right shoulder widths	54,024 (84%)	5,495 (9%)	4,286 (7%)	656 (1%)	163 (<1%)	64,624
Left shoulder width <4 Ft	8,769 (76%)	908 (8%)	1,365 (12%)	361 (3%)	123 (1%)	11,526
Other left shoulder widths	50,150 (84%)	5,089 (9%)	3,681 (6%)	498 (<1%)	107 (<1%)	59,525
Percent of trucks >20%	1,829 (71%)	164 (6%)	407 (16%)	134 (5%)	46 (2%)	2,580
Level roadway	56,179 (83%)	5,646 (8%)	4,632 (7%)	782 (1%)	203 (<1%)	67,442
Other roadway grade	2,502 (75%)	344 (10%)	400 (12%)	74 (2%)	26 (<1%)	3,346
Grand total	58,919 (83%)	5,997 (8%)	5,046 (7%)	859 (1%)	230 (<1%)	71,051

β_i: vector of estimable parameters

X_i: vector of explanatory (independent) variables (i.e., roadway, traffic, environmental characteristics, etc.,)

$EXP(\varepsilon_i)$: gamma-distributed error term

The error term, $EXP(\varepsilon_i)$ with a mean 1 and variance α, allows the variance to differ from the mean as shown in Equation 2 below (*17*):

$$VAR[y_i] = E[y_i] + \alpha E[y_i]^2 \tag{2}$$

Where:

$VAR[y_i]$: variance of crashes per year per segment

$E[y_i]$: mean of crashes per year per segment

α: over-dispersion parameter estimated with negative binomial model

A high α value indicates the presence of greater over-dispersion in the model, and as the over-dispersion parameter (α) approaches zero, the negative binomial model regresses to the Poisson model. The over-dispersion parameters for both crash frequency models developed as part of this study were statistically significant, indicating the appropriateness of the use of NB models. Ultimately, the model results are presented as the parameter estimates, β_i, for each explanatory variable, along with the standard error and *p*-value. In interpreting model results, negative β_i values represent an expected reduction in crash frequency, and positive values represent an expected increase in crash frequency. The results of the negative binomial models can give researchers a better understanding of how traffic, roadway, environmental, and other characteristics affect the expected frequency of crashes.

Crash Severity Analysis: Ordered Logit Model

Several past studies have successfully utilized discrete outcome models such as the ordered logit model in past traffic safety studies (*18*, *19*). The ordered logit model is often used for estimating the effect that explanatory variables have on the outcome of an ordered discrete variable: injury severity in this case of this study. The ordered logit model is derived by the unobserved variable, *Z*, which is used as the basis for modeling the ordinal ranking of data (*17*). The *Z* variable is specified as a linear function for each observation of occupant injury severity (*17*).

$$Z = \beta X + \varepsilon \tag{3}$$

Where:

X: vector of variables determining the discrete ordering for each occupant injury severity observation

β: vector of estimable parameters

ε: disturbance term

Considering this specification, the observed injury severity outcomes, *y*, is defined by the following thresholds:

$$\begin{aligned} y &= 1 \quad \text{if } z \leq \mu_0, \\ y &= 2 \quad \text{if } \mu_0 < z \leq \mu_1, \\ y &= 3 \quad \text{if } \mu_1 < z \leq \mu_2, \\ y &= 4 \quad \text{if } \mu_2 < z \leq \mu_3 \\ y &= 5 \quad \text{if } z > \mu_3, \end{aligned} \tag{4}$$

Where:

μ_i: estimable threshold parameters that define *y*, which corresponds to the ordered injury severity categories.

The thresholds, μ, are estimated along with the model parameters β_i. The first threshold (i.e., μ_0) is set to zero without loss of generality, and the error term, ε, is assumed to be logistically distributed across observations. Under this assumption, and by setting μ_0 equal to zero, the outcome probabilities become (*17*):

$$P(y = i) = F(u_i - \beta X) - F(u_{i-1} - \beta X) \tag{5}$$

Where:

F: Cumulative distribution function of the logistic distribution defining ε

μ_i: upper threshold for injury severity *i*

μ_{i-1}: lower Threshold for Injury severity *i*

Since each person is considered as a unique observation, panel data were incorporated into the severity models to account for potential correlation among the injury outcomes of persons in the same vehicle (i.e., potential intra-vehicle correlation). To achieve this framework, each occupant observation is assigned to a unique vehicle ID within the dataset. Further information regarding panel data is provided elsewhere (*17*).

Random Parameters

Given the variability of crash data, it was determined that random parameters would be considered in the negative binomial models and the ordered logit models. Random parameters have been proven to be effective in providing a better model fit when considering complex, unobserved, crash variables (*20*, *21*). The random parameters framework allows certain estimable parameters that exhibit significant variability (as evidenced by a significant standard deviation) to vary across observations. This also accounts for unobserved heterogeneity within the explanatory variables themselves. For example, there may be unobservable differences across the driving population such as risk-taking behavior and physiological factors (*17*). The random parameters framework alters both models as such:

$$\beta_i = \beta + \omega_n \tag{6}$$

where:

β_i: estimable parameter

ω_n: randomly distributed term (i.e., normally distributed with mean zero and variance σ^2)

Instead of randomly drawing parameters from their respective normal distributions, it is common practice to use Halton draws which accomplish the same result with far fewer draws (*22–24*). Due to a high number of observations and random parameters in the random parameter ordered logit models, 100 Halton draws were found to be adequate for this analysis.

In addition to the ordered logit model estimation, marginal effects were also estimated for additional insight in the severity analysis. The magnitude and sign of each marginal effect help to illustrate the effect that each variable has on severity outcomes (*17*). Because each variable in the random parameter ordered logit models were classified as a binary indicator, the numerical marginal effects represent the change in probability of an injury severity level when the corresponding indicator variable is changed from 0 to 1. Further information on the calculation of marginal effects is provided elsewhere (*17*). All models in this study were developed using the statistical software NLOGIT 5 (*25*).

Results and Discussion

The results of the random parameters negative binomial models (frequency models) are presented in Table 4. As described in the previous section, the random parameter framework allowed certain estimable parameters to vary across observations and this is represented in the results table with each random parameter having its own standard deviation, standard error, and *p*-value. Ultimately, the random parameters models provided a superior fit based on log likelihood ratio tests. The model results are interpreted such that a positive parameter, β, indicates that variable is associated with an increase in crash frequency for any given segment. For example, as the continuous variable, segment length, increases then the expected number of crashes at that segment would increase as well. Conversely, negative parameters represent an expected decrease in crash frequency. It is important to note that to have more uniform significant digits

Table 4. Results for the Random Parameter Negative Binomial Frequency Models

Freight-involved model	B	Std. error	*p*-value	Std. dev.	Std. error	*p*-value
Intercept	−7.138	0.852	<.001	0.113	0.017	<.001
Segment length*	0.184	0.008	<.001			
Ln(AADT)	0.704	0.073	<.001	0.014	0.002	<.001
Cable barrier*	0.180	0.102	.076			
Concrete barrier	0.237	0.089	.008	0.277	0.029	<.001
Median width < 39 (Ft)*	−0.181	0.058	.002			
Median width >80 (Ft)*	0.034	0.061	.575			
Right shoulder (Ft)*	0.002	0.011	.853			
Left shoulder (Ft)	−0.032	0.009	.001	0.007	0.002	.001
3 or 4 Lanes*	0.071	0.076	.347			
5 or 6 Lanes	−0.140	0.101	.167	0.324	0.035	<.001
Speed limit 65*	0.334	0.086	<.001			
Speed limit 75*	0.490	0.088	<.001			
Degree of curvature*	0.222	0.098	.023			
Percent grade	0.121	0.038	.002	0.137	0.019	<.001
Truck percent*	−0.006	0.004	.166			
Phoenix indicator*	0.437	0.095	<.001			
Over-dispersion**	0.117	1.101	<.001			
Non-freight model	β	Std. error	*p*-value	Std. dev.	Std. error	*p*-value
Intercept	−10.663	0.608	<.001	0.167	0.012	<.001
Segment length*	0.187	0.006	<.001			
Ln(AADT)	1.187	0.052	<.001	0.020	0.001	<.001
Cable barrier*	0.190	0.069	.006			
Concrete barrier	0.306	0.061	<.001	0.250	0.020	<.001
Median width <39 (Ft)*	−0.229	0.041	<.001			
Median width >80 (Ft)*	−0.267	0.043	<.001			
Right shoulder (Ft)*	−0.024	0.008	.004			
Left shoulder (Ft)	−0.016	0.007	.018	0.007	0.001	<.001
3 or 4 Lanes*	−0.132	0.052	.011			
5 or 6 Lanes	−0.312	0.071	<.001	0.266	0.025	<.001
Speed limit 65*	0.795	0.059	<.001			
Speed limit 75*	1.002	0.061	<.001			
Degree of curvature*	0.417	0.068	<.001			
Percent grade	0.133	0.026	<.001	0.222	0.013	<.001
Truck percent*	−0.018	0.003	<.001			
Phoenix indicator*	0.219	0.065	.001			
Over-dispersion**	0.128	0.445	<.001			

*Fixed Parameter in RP Model.

**Over-dispersion parameter for the negative binomial model framework.

Table 5. Results for the Freight-Involved Random Parameter Ordered Logit Severity Model

Variable	β	Std. error	*p*-value	Std. dev.	Std. error	*p*-value
Constant	−1.571	0.180	<.001	1.956	0.045	<.001
Summer months	−0.135	0.083	.101	1.227	0.071	<.001
Dust storm	1.113	0.267	<.001	2.282	0.258	<.001
Dark light conditions*	0.467	0.072	<.001			
Collision with concrete barrier	2.097	0.467	<.001	3.575	0.482	<.001
Rollover*	3.703	0.226	<.001			
Jackknife*	−2.120	0.914	.020			
Single vehicle*	−1.942	0.186	<.001			
Angle	−0.979	0.192	<.001	3.282	0.179	<.001
Head on	0.994	0.322	.002	2.694	0.329	<.001
Sideswipe same direction	−2.064	0.089	<.001	0.823	0.067	<.001
Sideswipe opposite direction	1.286	0.504	.011	1.381	0.544	.011
Motorcycle*	8.796	0.504	<.001			
Passenger vehicle*	1.949	0.081	<.001			
Cross median*	1.972	0.305	<.001			
Run-off-road right	2.164	0.128	<.001	2.149	0.112	<.001
Run-off-road left	2.426	0.119	<.001	1.264	0.094	<.001
Age 24 or less	−0.645	0.084	<.001	1.469	0.074	<.001
Age 65 or up	0.317	0.134	.018	1.429	0.120	<.001
Female*	0.740	0.068	<.001			
Safety device used*	−1.769	0.110	<.001			
Drugs or alcohol used	1.395	0.302	<.001	4.979	0.382	<.001
Median width <20 Ft*	−0.356	0.093	<.001			
Median width >80 Ft*	0.273	0.116	.019			
Speed limit 75	−0.345	0.112	.002	1.542	0.064	<.001
Right shoulder width <10 Ft	0.167	0.111	.133	0.263	0.078	.001
Left shoulder width <4 Ft	−0.836	0.123	<.001	1.524	0.069	<.001
Percent of trucks >20%	0.552	0.126	<.001	0.887	0.104	<.001
Level roadway*	−0.323	0.116	.006			
Threshold 1	1.379	0.039	<.001			
Threshold 2	5.003	0.110	<.001			
Threshold 3	7.264	0.198	<.001			
Restricted log likelihood (LL)	−6334.221					
Final LL for fixed model	−5872.399					
Final LL for RP model	−5820.096					

*Fixed parameter in RP model.

in the results, the natural log of AADT was used instead of AADT itself.

The results of the random parameter ordered logit models (severity models) and the associated marginal effects are presented in Tables 5 through 8. In total, 16 of the total 28 final variables included in the freight-involved random parameter model exhibited significant variability. During the modeling process, the variables with significant variability were retained as random. Variables that had statistically significant parameters but non-significant standard deviations were retained as fixed parameters. Variables that did not have significant parameters or significant variability were not included in the final models. Similar to the negative binomial model, the ordered logit model results are interpreted with positive parameters indicating an increase in the probability of the most severe injury severity outcome, and vice versa for negative parameter estimates.

While the marginal effects of several variables are small, when the costs of severe motor vehicle crashes (*5*) are considered, even a slight increase in the risk of severe injury can cost society millions of dollars over the course of several years. It is also important to note that these effects are compounded over the entire I-10 corridor through Arizona and thus many small risks can become one big problem over time.

Negative Binomial Frequency Model Results Discussion

Length and AADT had positive effects for both freight and non-freight crashes (i.e., greater segment lengths and AADT increased crash frequency) which is both intuitive and well supported by many past traffic safety studies. These two variables are also the standard for developing base safety performance functions (SPFs) and the parameter estimates and constant terms presented in both

Table 6. Marginal Effects for the Freight-Involved RP Ordered Logit Model

Variable	1-no injury	2-injury	3-injury	4-injury	5-fatal
Summer months	0.00337*	−0.00249*	−0.00086*	−0.10013	−0.10014
Dust storm	−0.04936***	0.03593***	0.01307**	0.00003**	0.00004**
Dark light conditions	−0.01345***	0.00991***	0.00345***	0.00009***	0.00001***
Collision with concrete barrier	−0.15382**	0.10816**	0.04436**	0.00116*	0.00013*
Rollover	−0.47921***	0.28201***	0.19059***	0.00591***	0.00069***
Jackknife	0.02333***	−0.01732***	−0.00585***	−0.00014***	−0.67795
Single vehicle	0.02656***	−0.01969***	−0.00669***	−0.00017***	−0.77646
Angle	0.01700***	−0.01260***	−0.00429***	−0.00011***	−0.4982
Head on	−0.04171**	0.03043**	0.01098**	0.00028*	0.00003*
Sideswipe same direction	0.05104***	−0.03752***	−0.01316***	−0.00033***	−1.54131
Sideswipe opposite direction	−0.06297	0.04564	0.01685	0.00043	0.00005
Motorcycle	−0.96824***	−0.00315	0.42783***	0.43308***	0.11049**
Passenger vehicle	−0.06161***	0.04507***	0.01609***	0.00040***	0.00005***
Cross median	−0.13571***	0.09606***	0.03854***	0.00100***	0.00012***
Run-off-road right	−0.14844***	0.10480***	0.04241***	0.00110***	0.00013***
Run-off-road left	−0.18320***	0.12777***	0.05385***	0.00142***	0.00017***
Age 24 or less	0.01415***	−0.01046***	−0.00359***	−0.41797	−0.418
Age 65 or up	−0.00934**	0.00688**	0.00240**	0.00006**	0.000007**
Female	−0.02231***	0.01641***	0.00574***	0.00014***	0.00002***
Safety device used	0.10238***	−0.07339***	−0.02819***	−0.00072***	−3.38554
Drugs or alcohol used	−0.07173***	0.05186***	0.01933**	0.00049**	0.00006**
Median width <20 Ft	0.00833***	−0.00615***	−0.00212***	−0.24657	−0.24659
Median width >80 Ft	−0.00777**	0.00572**	0.00199**	0.00005**	.000006**
Speed limit 75	0.00835***	−0.00616***	−0.00212***	−0.24751	−0.24753
Right shoulder width <10 Ft	−0.00456	0.00336	0.00117	0.00003	0.000003
Left shoulder width <4 Ft	0.01834***	−0.01355***	−0.00466***	−0.00012***	−0.54237
Percent of trucks >20%	−0.01782***	0.01309***	0.00460***	0.00011***	0.00001***
Level roadway	0.02163***	−0.01587***	−0.00560***	−0.00014***	−0.65657

Note: ***, **, * = significance at 1%, 5%, 10% levels.

frequency models could be used for various future safety studies on the I-10 involving freight or non-freight type crashes.

Segments with either concrete median barriers or cable median barriers tend to experience more freight and non-freight crashes than segments with unprotected medians. This is consistent with past research (*26*) because when a vehicle runs off the road and collides with a median barrier the crash is nearly always reported. However, if a vehicle runs off the road in a rural area with no median barrier it is possible that the incident may not be reported if the vehicle does not collide with another object. While the presence of median barriers may appear to be a safety hazard, it has also been proven in past studies that median barriers are effective in decreasing the severity of run-off-road crashes (*26*). For this study as well, higher crash frequencies being associated with the presence of median barriers is likely a result of median barriers existing in areas that are already prone to high crash frequencies. Interestingly, the presence of median barriers appears to have very similar effects for freight crashes and non-freight crashes.

Segments with low median widths have a similar effect for both freight and non-freight crashes where they correspond with a decrease in crashes which is contradictory to past research (*26*, *27*). This may be due to a long stretch of wide medians with barriers through some of the most heavily traveled segments in Phoenix. Many of the areas with small median widths are in the suburban sections of Phoenix that experience relatively lower AADT compared with central Phoenix. Segments with high median widths are not significant for freight crashes and also indicate a decrease in frequency of non-freight crashes which is in alignment with past research (*27*). These results may be an indicator that the I-10 is adequately designed in terms of median width.

The continuous variables right shoulder width and left shoulder width all indicated that an increase in width decreases the frequency of crashes which is consistent with past studies (*28*). The one exception seen in these results is that the effect of right shoulder width is not significant for freight crashes. This may be because freight vehicles are larger and perhaps more top heavy, the difference between a 10-foot shoulder and a 12-foot shoulder for example may not make a significant difference in crash reductions. Also, there is not much variability in the left shoulder width on the I-10 with the

Table 7. Results for the Non-Freight Random Parameter Ordered Logit Severity Model

Variable	β	Std. error	p-value	Std. dev.	Std. error	p-value
Constant	−1.975	0.155	<.001	3.115	0.026	<.001
Summer months	0.010	0.034	.780	0.188	0.030	<.001
Dust storm	−0.513	0.390	.189	3.536	0.354	<.001
Dark light conditions*	0.332	0.034	<.001			
Collision with concrete barrier	2.252	0.088	<.001	1.240	0.072	<.001
Rollover*	4.505	0.074	<.001			
Jackknife*	−3.032	0.546	<.001			
Single vehicle*	−1.408	0.058	<.001			
Angle	−0.076	0.100	.444	2.079	0.095	<.001
Head on	1.265	0.201	<.001	3.828	0.201	<.001
Sideswipe same direction	−1.580	0.053	<.001	1.249	0.047	<.001
Sideswipe opposite direction	0.326	0.328	.320	0.517	0.336	.124
Motorcycle*	6.956	0.167	<.001			
Passenger vehicle*	1.458	0.131	<.001			
Cross median*	1.889	0.155	<.001			
Run-off-road right	2.016	0.059	<.001	2.102	0.049	<.001
Run-off-road left	2.020	0.056	<.001	0.954	0.042	<.001
Age 24 or less	−0.727	0.033	<.001	0.897	0.027	<.001
Age 65 or up	0.246	0.063	<.001	0.830	0.060	<.001
Female*	0.822	0.030	<.001			
Safety device used*	−2.745	0.063	<.001			
Drugs or alcohol used	1.732	0.127	<.001	4.488	0.144	<.001
Median width <20 Ft*	−0.145	0.036	<.001			
Median width >80 Ft*	0.256	0.061	<.001			
Speed limit 75	−0.573	0.056	<.001	0.906	0.034	<.001
Right shoulder width <10 Ft	−0.040	0.058	.491	0.635	0.045	<.001
Left shoulder width <4 Ft	−0.120	0.058	.038	1.197	0.035	<.001
Percent of trucks >20%	0.302	0.080	<.001	1.145	0.068	<.001
Level roadway*	−0.323	0.116	.006			
Threshold 1	1.920	0.022	<.001			
Threshold 2	6.064	0.057	<.001			
Threshold 3	9.413	0.121	<.001			
Restricted log likelihood (LL)	−35,695.375					
Final LL for fixed model	−33,150.964					
Final LL for RP model	−33,088.906					

*Fixed parameter in RP model.

mean being 10.64 Ft and the standard deviation being 1.93 Ft.

Speed indicators for segments with 65 mph and 75 mph speed limits have the same positive effect for both freight and non-freight crashes and are significant in both models (compared with segments with 45–55 mph speed limits). In previous studies, results for the effect of speed limit on the frequency of total crashes are mixed; however, studies analyzing the severity of crashes often conclude that higher speed limits are often correlated with high fatality rates (*29*). It is interesting that freight crashes are not affected differently than non-freight crashes. It seems intuitive that for the I-10, segments with 65 mph speed limits would have more crashes due to their proximity to urban areas. This result may be due to the conflicts created by speed differentials between freight and passenger vehicles on rural segments.

High degrees of curvature and high percent grade prove to be positive factors for both freight and non-freight type crash frequencies. This result is consistent among most traffic safety studies (*30*). Drivers may enter curves and high grades at unsafe speeds which in turn, leads to higher crash frequencies.

The continuous variable for truck percent is interesting in that its negative effect on crash frequency is not significant for freight crashes but it is for non-freight crashes. This may be an indicator that passenger vehicles are often the ones at fault in truck crashes. However, it could also be another function of vehicle exposure. Low truck percentages are most often observed on urban segments with high passenger vehicle volumes. An Ontario, Canada study found the same result to be true (*9*). The "Phoenix" indicator variable represents the urban area through Phoenix and it has a significant positive effect for both freight and non-freight crashes. This result was

Table 8. Marginal Effects for the Non-Freight RP Ordered Logit Model

Variable	1-no Injury	2-injury	3-injury	4-injury	5-fatal
Summer months	−0.00021	0.00017	0.00003	0.00001	0.00000
Dust storm	0.00868*	−0.00074*	−0.00129*	−0.34517	−0.34518
Dark light conditions	−0.00778***	0.00659***	0.00117***	0.00001***	0.0000006***
Collision with concrete barrier	−0.14684***	0.12128***	0.02515***	0.00040***	.00014***
Rollover	−0.61178***	0.41465***	0.19322***	0.00377***	0.00014***
Jackknife	0.02092***	−0.01779***	−0.00308***	−0.82162	−0.82161
Single vehicle	0.02084***	−0.01769***	−0.00310***	−0.8282	−0.82822
Angle	0.00158	−0.00134	−0.00024	−0.06325	−0.06325
Head on	−0.05131***	0.04314***	0.00804***	0.00013***	0.000005***
Sideswipe same direction	0.02219***	−0.01884***	−0.00330***	−0.88158	−0.88159
Sideswipe opposite direction	−0.00817	0.00691	0.00123	0.00002	0.000001
Motorcycle	−0.93613***	0.17474***	0.71247***	0.04711***	0.00180***
Passenger vehicle	−0.01749***	0.01486***	0.00258***	0.00004***	0.000001***
Cross median	−0.10623***	0.08843***	0.01752***	0.00028***	0.00001***
Run-off-road right	−0.10990***	0.09149***	0.01811***	0.00029***	0.00001***
Run-off-road left	−0.10839***	0.09028***	0.01782***	0.00028***	0.00001***
Age 24 or less	0.01410***	−0.01196***	−0.00211***	−0.56421	−0.56423
Age 65 or up	−0.00587***	0.00497***	0.00088***	0.00001***	0.000001***
Female	−0.01875***	0.01589***	0.00282***	0.00004***	0.000002***
Safety device used	0.22251***	−0.18066***	−0.04115***	−0.00067***	−11.50761
Drugs or alcohol used	−0.08910***	0.07441***	0.01445***	0.00023***	0.000008***
Median width <20 Ft	0.00298***	−0.00253***	−0.00045***	−0.11944	−0.11944
Median width >80 Ft	−0.00611***	0.00518***	0.00092***	0.00001***	0.000001***
Speed limit 75	0.01031***	−0.00875***	−0.00154***	−0.41095	−0.41096
Right shoulder width <10 Ft	0.00084	−0.00072	−0.00013	−0.03385	0.00000
Left shoulder width <4 Ft	0.00247**	−0.00210**	−0.00037**	−0.09893	−0.09894
Percent of trucks >20%	−0.00742***	0.00628***	0.00112***	0.00002***	0.000001***
Level roadway	0.02447***	−0.02067***	−0.00374***	−1.00502	−1.00512

Note: ***, **, * = significance at 1%, 5%, 10% levels.

to be expected: during the data collection process it became evident that a large portion of the sample crashes occurred within the Phoenix area and this is largely the result of high congestion and other potential unobserved characteristics associated with highly populated urban areas.

Ordered Logit Severity Model Results Discussion

One of the most important findings in the severity analysis was that the weather variable, blowing sand and/or dust had a significant effect on the severity of freight-involved crashes as opposed to other weather variables, such as rain, which did not have a significant effect on the severity of freight or non-freight crashes. This finding is of particular significance in Arizona and the southwest USA in general as the stretch of I-10 between Phoenix and Tucson experiences dust storms on a frequent basis, which creates hazardous driving conditions. Dust-storm related crashes are over represented in the freight dataset (i.e., 1.5% of all freight-involved records and 0.3% of non-freight records). This over representation may be due to a more substantial decrease in visibility during dust storms for freight vehicles. Because a freight vehicle's field of vision is higher than a normal passenger vehicle, they may be more at risk of colliding with small vehicles or other objects close to the ground. Trucking companies should make efforts to educate their drivers on what steps to take in the event of a dust storm.

None of the seasons (i.e., winter, summer, etc.) were significant in either severity model, however, summer did display significant variability and therefore was left in the model as a random parameter. In many states with harsh winters, studies have shown that crash severity is often reduced during winter months potentially due to more cautious driving behavior in adverse winter weather conditions (*31*), however this is not the case on the I-10 in Arizona. Another interesting finding in the severity model is that crashes occurring during dark lighting conditions resulted in more severe outcomes for both freight and non-freight crashes. This is likely due to deteriorating visual capabilities and driving behavior at later times in the day and lack of street lighting in rural areas. Dark lighting conditions were also seen to increase injury severity for large truck crashes across the entire interstate system (*32*).

Several indicator variables were created based on the first harmful event recorded for each crash-involved vehicle. While most did not have a statistically significant impact in either severity model, several of them did. For example, crashes with the first harmful event recorded as a rollover resulted in the second highest injury severity outcomes for both freight and non-freight. Collisions with concrete barriers also resulted in more severe injury outcomes. Crashes that had a first harmful event coding of jackknife actually resulted in less severe crashes than other first harmful events. This finding agrees with a study completed in 2014 that found that jackknife crashes were less severe when compared to rollover crashes (*33*).

With respect to collision manner, crashes that involved just a single vehicle were less severe for both freight and non-freight crashes. This is often the case on high speed controlled access highways where there are few hazardous obstructions near the road. Head-on collisions were significant with a positive effect for both freight and non-freight crashes. This is consistent to one past study done in Canada which showed that head-on collisions with large trucks resulted in more severe crashes (*34*).

Injuries were more severe for people in passenger vehicles as opposed to those in freight or other vehicles (a variety of other miscellaneous vehicle body styles such as garbage truck, dump truck, ambulance, etc...) for both freight and non-freight crashes. This result is both intuitive and consistent with past studies (*34*). Crashes involving motorcycles resulted in the most severe results out of any indicator in both the freight and non-freight models. Next, incidents where a vehicle ran off the road left, right, or crossed the median significantly resulted in more severe crashes. The severity of these crashes can be greatly affected by roadway and environmental conditions but in general they tend to result in more severe crashes.

Person-level variables revealed that female motorists were more likely to be injured than other gender identifiers. Also, motorists age 24 or younger were less likely to be injured. These results may be due to physiological differences among gender and age groups. Motorists who had reportedly used drugs or alcohol were more likely to be injured. This result is also common in traffic studies and it has become a large educational campaign across the USA. Another common result was that motorists who used a safety device (i.e., lap belt, shoulder and lap belt, and helmet) were less likely to be injured. These results were significant and had the same effect for both freight and non-freight crashes.

For both freight and non-freight, crashes on segments with low median widths (less than 20 Ft) were less severe while crashes on segments with high median widths (more than 80 Ft) were more severe. Many of the segments with low median widths also had median barriers to prevent head-on collisions and were in urban areas where congestion-related rear-end crashes are common. Interestingly, crashes on segments with 75 mph speed limits were less severe than those with 65 mph or 55 mph speed limits for non-freight, however, this variable was not significant for freight crashes. Narrow right shoulder widths (less than 10 Ft) and narrow left shoulder widths (less than 4 Ft) had significant effects for freight crashes but not non-freight crashes. Narrow right shoulder widths resulted in more severe crashes and narrow left shoulder widths resulted in less severe crashes. Low shoulder widths often led to higher frequencies and higher severity crashes. In this case, the decrease in severity with narrow left shoulder width might again be a matter of urban crashes versus rural crashes. Segments with high percent truck volume experienced more severe outcomes for freight crashes but were not significant for non-freight crashes. Crashes on level roads experienced less severe crashes for both freight and non-freight as compared to crashes on downhill or uphill segments. Interestingly, crashes on curves were not significant in terms of injury severity for either model.

Conclusion

The I-10 corridor through Arizona is of significant importance to goods movement throughout the Southwest and thus its operation is essential to the well-being of the general public and U.S. businesses. Recent studies on the efficiency and safety of the I-10 have revealed that many sections are areas of concern with low safety indexes. Additionally, studies on the safety of freight transport and how it relates to non-freight type transport are few and far between.

Two different statistical methods (RP negative binomial regression and RP ordered logit regression) were used to analyze factors which may lead to high crash frequencies and severe injury outcomes on the I-10 through Arizona. The results of the frequency models revealed that geometric characteristics such as median width, shoulder width and number of lanes were generally less significant for freight-involved crashes than non-freight crashes. Another interesting finding in the frequency models was that both high (i.e., $>$80 Ft) and low (i.e., $<$39 Ft) median widths indicated a decrease in crash frequency for non-freight crashes. The results of the severity models revealed that dust-storm crashes were significant and had a positive effect on the severity of freight-involved crashes but they were not significant for non-freight crashes. Head on crashes were not significant for the injury severity of freight-involved crashes but they were significant with a positive effect for the injury severity in non-freight crashes.

In order to improve safety and efficiency on a major freight corridor freeway, agencies should focus on education and enforcement in addition to employing new ITS solutions as opposed to widening shoulders, adding lanes, or changing speed limits. Variable speed limits may help to decrease the severity of dust-storm-related incidents. Dynamic lane control could help reduce the frequency of sideswipe and rear-end crashes in the often congested urban areas of Phoenix and Tucson. Driver-assistance technology for freight vehicles, such as lane departure and collision warning systems, could reduce the frequency and severity of run-off-road, single vehicle, and rollover crashes.

Further studies on the frequency and severity of freight vehicle crashes on either urban or rural areas only, would be beneficial in understanding unobserved spatial factors. For example, fatigue is likely a larger factor in rural areas as opposed to urban areas and the safety impact of wider left shoulder widths might be different in rural areas as opposed to urban areas.

Acknowledgments

Funding for this research was provided in part by the Arizona Board of Regents (ABOR), and data were obtained from ADOT.

Author Contributions

The authors confirm contribution to the paper as follows: study conception and design: Brendan Russo, Samuel Taylor, and Emmanuel James; data collection: Brendan Russo, Samuel Taylor, and Emmanuel James; analysis and interpretation of results: Brendan Russo and Samuel Taylor; draft manuscript preparation: Brendan Russo and Samuel Taylor. All authors reviewed the results and approved the final version of the manuscript.

References

1. U.S. Department of Transportation. *Economic Impact of Freight*. https://www.rita.dot.gov/bts/programs/freight_-transportation/html/freight_and_growth.html. Accessed July 5, 2017.
2. U.S. Department of Transportation. *Freight Facts and Figures 2015*, 11th ed.https://www.rita.dot.gov/bts/sites/rita.-dot.gov.bts/files/FFF_complete.pdf. Accessed July 5, 2017.
3. Statistics Department National Safety Council. *NSC Motor Vehicle Fatality Estimates*. http://www.nsc.org/NewsDocuments/2017/12-month-estimates.pdf. Accessed July 5, 2017.
4. U.S. Department of Transportation. *Pocket Guide to Large Truck and Bus Statistics*. June 2017. https://ntl.bts.gov/lib/61000/61900/61913/2017-Pocket-Guide-to-Large-Truck-and-Bus-Statistics-Final-508C-01.2.pdf. Accessed July 6, 2017.
5. National Safety Council. *Estimating the Costs of Unintentional Injuries*. March 2015. http://www.nsc.org/NSCDocuments_Corporate/estimating-costs.pdf. Accessed July 5, 2017.
6. I-95 Corridor Coalition. *The Coalition*. http://i95coalition.org/the-coalition-2/. Accessed July 14, 2017.
7. I-10 Corridor Coalition. *Interstate 10 Connects People, Businesses, States and More*. https://i10connects.com/. Accessed July 14, 2017.
8. HDR, ADOT. *I-10 Corridor Profile Study SR 202L to New Mexico State Line*. https://www.azdot.gov/docs/default-source/planning/Corridor-Studies/cps3-i-10e-draft-wp1-(jan-2016).pdf?sfvrsn=2. Accessed July 3, 2017.
9. Ran, R. *Prediction of Frequencies of Truck-Involved and Non-Truck-Involved Crashes on Roadway Segments in Ontario*. University of Windsor, October 2015.
10. Dong, C., Q. Dong, H. Huang, W. Hu, and S. S. Nambsian. Estimating Factors Contributing to Frequency and Severity of Large Truck-Involved Crashes. *Journal of Transportation Engineering*, Vol. 143, No. 8, 2017. https://ascelibrary.org/doi/10.1061/JTEPBS.0000060
11. AECOM, ADOT. *I-10/SR 85 Corridor Profile Study*. https://www.azdot.gov/docs/default-source/planning/Corridor-Studies/i-10-sr85-executive-summary.pdf?sfvrsn=2. Accessed July 3, 2017.
12. ESRI 2011. *ArcGIS Desktop: Release 10*. Environmental Systems Research Institute, Redlands, CA, 2011.
13. Arizona Department of Transportation (ADOT). *Arizona's Crash Report Forms Instruction Manual*, 9th ed., pp.24–25, Arizona Department of Transportation, Phoenix, AZ, 2010.
14. Mohammadi, M., V. A. Samaranayake, and G. Bham. Crash Frequency Modeling Using Negative Binomial Models: An Application of Generalized Estimating Equation to Longitudinal Data. *Analytic Methods in Accident Research*, Vol. 2, 2014, pp. 52–69.
15. Jung, S., S. Joo, and O. Cheol. Evaluating the Effects of Supplemental Rest Areas on Freeway Crashes Caused by Drowsy Driving. *Accident Analysis and Prevention*, Vol. 99, 2017, pp. 356–363.
16. Srinivasan, R., and K. Bauer. *Safety Performance Function Development Guide: Developing Jurisdiction Specific SPFs*. Publication FHWA-SA-14-005. FHWA, September 2013.
17. Washington, S., M. Karlaftis, and F. Mannering. *Statistical and Econometric Methods for Transportation Data Analysis*, 2nd ed. Chapman and Hall/CRC, Boca Raton, FL, 2011.
18. Bogue, S., R. Pateli, and L. Balan. A Modified Rank Ordered Logit Model to Analyze Injury Severity of Occupants in Multivehicle Crashes. *Analytic Methods in Accident Research*, Vol. 14, 2017, pp. 22–40.
19. Osman, M., R. Paleti, S. Mishra, and M. M. Golias. Analysis of Injury Severity of Large Truck Crashes in Work Zones. *Accident Analysis & Prevention*, Vol. 97, 2016, pp. 261–273.
20. Monatana Department of Transportation. *Safety and Operational Impacts of Differential Speed Limits on Two-Lane Rural Highways in Montana*. Publication FHWA/MT-16006/8224-001, 2016.
21. Quddus, M. Effects of Geodemographic Profiles of Drivers on Their Injury Severity from Traffic Crashes Using Multilevel Mixed-Effects Ordered Logit Model. *Transportation*

Research Record, Vol. 2514, 2015, pp. 149–157. https://doi.org/10.3141/2514-16
22. Train, K. *Halton Sequences for Mixed Logit*. University of California, Berkeley Department of Economics, July 22, 1999.
23. Zeng, T. Using Halton Sequences in Random Parameters Logit Models. *Journal of Statistical and Economic Methods*, Vol. 5, No. 1, 2016, pp. 59–86.
24. Hensher, D. A., J. M. Rose, and W. H. Greene. *Applied Choice Analysis: A Primer*. Cambridge University Press, New York, 2005.
25. NLOGIT 5. Econometric Software Inc., 2012.
26. Savolainen, P., T. Gates, B. Russo, and J. Kay. Study of High-Tension Cable Barriers on Michigan Roadways MDOT ORBP Project Number: OR10-036, October 2014.
27. The Association of Median Width and Highway Accident Rate. Publication FHWA-RD-93-046, August 1993. https://www.fhwa.dot.gov/publications/research/safety/humanfac/93046/index.cfm. Accessed July 21, 2017.
28. Park, J., and M. Abdel-Aty. Evaluation of Safety Effectiveness of Multiple Cross Sectional Features on Urban Arterials. *Accident Analysis & Prevention*, Vol. 92, 2016, pp. 245–255.
29. Savolainen, P., T. Gates, E. Hacker, A. Davis, S. Frazier, B. Russo, E. Rista, M. Parker, F. Mannering, and W. Schneider. Evaluating the Impacts of Speed Limit Policy Alternatives. *MDOT Research Administration*. Project Number: OR 13-009, July 2014.
30. Gooch, J., V. Gayah, and E. Donnell. Quantifying the Safety Effects of Horizontal Curves on Two-Way, Two-Lane Rural Roads. *Accident Analysis & Prevention*, Vol. 92, 2016, pp. 71–81.
31. Russo, B., P. Savolainen, W. Schneider, and P. Anastasopoulos. Comparison of Factors Affecting Injury Severity in Angle Collisions by Fault Status Using a Random Parameters Bivariate Ordered Probit Model. *Analytic Methods in Accident Research*, Vol. 2, 2014, pp. 21–29.
32. Islam, M., and S. Hernandez. Large Truck-Involved Crashes: Exploratory Injury Severity Analysis. *Journal of Transportation Engineering*, Vol. 139, No. 6, 2013, pp. 596–604.
33. Lemp, J. D., K. M. Kockelman, and A. Unnikrishnan. Analysis of Large Trucks Crash Severity Using Heteroskedastic Ordered Probit Models. *Accident Analysis and Prevention*, Vol. 43, No. 1, 2011, pp. 370–380.
34. Li, X. Analysis of Injury Severity of Drivers Involved in Single-Vehicle and Two-Vehicle Crashes on Ontario Highways. *University of Windsor*, 2014.

The Standing Committee on Truck and Bus Safety (ANB70) peer-reviewed this paper (18-02290).

The opinions, findings, and conclusions expressed in this manuscript are those of the authors and not necessarily those of ABOR or ADOT.

Article

TRR
JOURNAL OF THE TRANSPORTATION RESEARCH BOARD

Transportation Research Record
2018, Vol. 2672(34) 63–72

DOI: 10.1177/0361198118786672
journals.sagepub.com/home/trr
SAGE

Uniformity of Terminology for Circular Intersection Designs

Joe Gustafson[1]

Abstract

Most transportation engineers around the world, and now in the United States, are relatively familiar with roundabouts and their operational and safety benefits. Although roundabouts are becoming increasingly common, drivers and even engineering professionals often contend with mixed messages about roundabout design and operation. In a world speckled with all manner of spiral roundabouts, signalized roundabouts, traffic circles, gyratories, and rotaries, is it any wonder that confusion, and public resistance, often persists? These mixed messages may represent the greatest hurdle to implementation, public acceptance, and safe operation of multi-lane roundabouts in particular. Within North America and across the globe, circular intersection designs that appear relatively similar to users can in fact require significantly different driver behaviors, depending on whether they are configured with a continuous circle road or a network of crossing roadways. This distinction can be of critical importance for roadway designers and agencies, elected officials and other policymakers, road user education and licensing, traffic enforcement, mapping and GPS navigation, and safe operation of autonomous vehicles. This paper aims to provide an overview of existing definitions, explore the nature of conflict points for each design, provide a framework modeling method for analysis, and provide globally applicable definitions for roundabout features for use in design, education, policy, enforcement, and research. This paper is focused primarily on roundabout design guidance and operations within the United States, but places these practices within the global context, such that the definitions and analyses provided can be applied to all forms of roundabout intersections around the world.

Variations in roundabout design, specifically lane configuration and traffic controls, can have significant implications for expected driver behaviors. More importantly, many such variations in design can and do exist among roundabout intersections across the world and even within the same communities. In an interconnected society it is likely that many drivers will have encountered, in person or in media, a roundabout design that elicits a set of driver behaviors which may be incorrect or even hazardous at other roundabouts containing different design features.

Unfortunately, no widely accepted set of terminology appears to exist that would differentiate between these various designs, despite their markedly different operational characteristics. Even at single-lane roundabouts, differences in signing, striping, and traffic control can shape driver expectations and behaviors that will subsequently be expressed when the same driver encounters a multi-lane design.

For the purpose of differentiating various roundabout designs, this paper will make considerable use of the term "carriageway," widely used in the United Kingdom. Whereas the terms "highway" or "roadway" can be interpreted to apply jointly to both sides of a divided highway (note the singular form), "carriageway" refers specifically to a contiguous group of lanes which is spatially or physically separated from other contiguous lane groups, such as opposing lanes or channelized turn lanes.

Existing definitions of roundabouts in design guidelines and other sources often contain common themes, most notably traffic flow in one direction around a central island. However, existing definitions generally fail to differentiate between design types that operate on fundamentally different operational principles, or fail to include broad categories of intersection types that are considered by the public or other practitioners to also be "roundabouts."

For example, the 2009 edition of the Manual on Uniform Traffic Control Devices (MUTCD), published by the United States Federal Highway Administration (FHWA) with input from various American professional associations, provides the following definition of roundabouts:

> A circular intersection with yield control at entry, which permits a vehicle on the circulatory roadway to proceed, and with deflection of the approaching vehicle counterclockwise around a central island. (*1*)

[1]Traffic Engineer, Washington County, Stillwater, MN

Corresponding Author:
Address correspondence to Joe Gustafson:
joe.gustafson@co.washington.mn.us

Another important source of guidance, specifically focused on roundabouts and widely used by professionals in the United States, is "Roundabouts: An Informational Guide" 2nd edition, also known as National Cooperative Highway Research Program (NCHRP) Report No. 672, hereinafter referred to simply as "NCHRP 672." NCHRP 672 provides the following definition of roundabouts:

> Roundabouts are a subset of circular intersections with specific design and traffic control features. These features include yield control of all entering traffic, channelized approaches, and geometric curvature and features to induce desirable vehicular speeds. (*2*)

Although these documents were developed for an audience of American practitioners, it is notable that these definitions both contain three primary themes:

- Circular shape
- Deflection or geometric curvature
- Yield control at entry

Although concise and largely in keeping with best practices for design of many roundabouts, these criteria can exclude many inplace designs that are nevertheless considered by many to be roundabouts. For example, an intersection may include a central island which is not truly "circular" but may in fact be shaped as a teardrop, "dog-bone," ellipse, or rounded rectangle. Approach deflection, although desirable to reduce vehicle speeds, is absent from many designs, especially "turbo" roundabouts where approach roadways often intersect a circular roadway at approximately 90°. Furthermore, these definitions both indicate that an intersection with similar geometry but controlled or regulated by traffic signals would in fact not be considered to be a roundabout at all, despite such junctions being fairly common across the globe and despite them often being considered to be a form of roundabout.

These definitions also make no mention of the operational configuration or driver behavior when the roundabout contains multiple lanes. For that, another important source of American guidance to consider is that which applies largely to drivers and policymakers rather than to roadway designers: The Uniform Vehicle Code (UVC). Although the UVC has not been updated since the year 2000, it remains the basis for inplace traffic operations laws across the U.S. Notably, the UVC does not contain any references specific to roundabouts (*3*), and, accordingly, the same is largely true of state and local traffic laws in the U.S. as well. Therefore, driver behavior at roundabouts is, understandably, governed by the same laws that were written to apply widely across the driving environment, rather than by a unique set of driving laws specific to roundabouts. Drivers are seldom familiar with the specific verbiage of traffic laws but are familiar with the general "rules of the road," and will navigate roundabouts according to these same behaviors and norms to which they are already accustomed, such as lane discipline, signaling, turning, and yielding.

It is imperative to consider the full universe of "roundabout" designs, including those which are not necessarily circular, junctions of all sizes and operating speeds, and all forms of traffic control including traffic signals and free-flow (merging) designs. The unifying feature that remains within this universe of roundabout designs is the presence of a central island around which traffic progresses in only one direction. Beyond this unifying feature, an important distinction separates roundabout designs into one of two groups, defined below:

> **Concentric roundabout:** A junction formed by a series of three-legged ("T") intersections with a continuous roadway encircling a central island.

> **Crossing roundabout:** A junction between two roadways in which turns made across opposing traffic are completed by way of the far side of a central island.

Concentric roundabouts include many inplace roundabouts in Continental Europe, including "turbo" roundabouts as constructed in the Netherlands. A defining characteristic of a concentric design is that a driver executes a *turn* to enter a circular roadway and another turn to exit the intersection. A *turn* should not be confused with simple steering to remain within a curved lane; rather, a *turn* constitutes a lane departure from one lane onto a separate intersecting lane. A driver within the circulatory roadway of a concentric roundabout who does not execute a turn will circulate endlessly unless a turn movement is forced, as is the case with turbo roundabouts which use physical channelization to force turns to be made from the circulatory roadway. Even at a single-lane roundabout, a concentric design can often be readily identified by lane striping across the roundabout exits, as such striping demarks the outer edge of the circulating lane, which a driver departs to reach the exit roadway.

Crossing roundabouts, by comparison, encompass most roundabout designs in North America, Australia, and the United Kingdom, though certainly not exclusively because wide variation in designs often exists. As shown in Figure 3 later in this paper, a defining characteristic of a crossing design is that the through lanes lead *out* of the roundabout, rather around the center island in a circular loop. All examples of roundabouts illustrated within the 2009 MUTCD, including single-lane roundabouts, are crossing designs, as evidenced by the choice of pavement arrows, the lane configuration, and the lack of lane striping across exit paths. Even for multi-lane roundabouts which do not contain striping within the inscribed circle area, crossing designs can be identified based on the required behaviors for exiting and circulating traffic.

Both concentric and crossing roundabouts can offer improved safety over conventional intersection types, and

Figure 1. Multi-lane concentric roundabout, consisting of inner and outer ring lanes in Wrotclaw, Poland.

both are widely used. Concentric roundabouts may be more naturally grasped by drivers, and may also align with cultural expectations based on portrayal in media. However, concentric multi-lane roundabouts (such as that shown in Figure 1) can create difficulties in exiting from their inner ring, because traffic circulating in their outer ring can block the exit path, and it becomes incumbent on the driver in the inner lane to first select a gap in the outer ring in order to exit from the circulatory roadway. The inner ring therefore becomes underutilized, and capacity is therefore reduced (*4*). The primary advantage of crossing roundabouts over concentric roundabouts is the ability to carry two or more lanes of traffic through the roundabout, by allowing exits to occur from all lanes without the need to first change lanes into an outer ring. Turbo roundabouts can allow for better usage of the inner lane of concentric roundabouts, but may necessitate less desirable entry angles and speed differentials.

Much as a divided highway may be considered "one" highway consisting of two separate carriageways, a roundabout may similarly be considered to be "one" junction consisting of several separate conflict locations. The individual vehicular conflict locations that occur within roundabouts, and the associated driver behaviors at each conflict location, are fully analogous to those occurring at junctions between one-way roadways. However, the arrangement of these conflict points, and the required driver behaviors at each, differ significantly between concentric and crossing roundabout designs, particularly where multiple lanes are present.

Understanding Conflict Points

One explanation frequently offered to support the safety and simplicity of roundabout intersections is that they feature fewer "conflict points" than conventional intersections, and that the conflicts that do occur at roundabouts are less hazardous than those of conventional intersections. Although it is true that roundabouts have fewer, and safer, conflict points than conventional intersections, crossing roundabouts have a different number of conflict points and types than concentric roundabouts. Failure to correctly understand these conflict types can lead to misleading public outreach efforts, signing and striping errors, and higher crash rates. Therefore, it can be particularly useful for a roundabout designer or operating agency to properly distill the conflict points of a given roundabout design to reduce these errors.

Conflict points are described in NCHRP 672 as "A location where the paths of two vehicles, or a vehicle and a bicycle or pedestrian, merge, diverge, cross, or queue behind each other" (*2*). The report goes on to describe three types of conflicts, as follows:

- Crossing conflict: The intersection of two traffic streams, including pedestrians. Crossing conflicts are the most severe type of conflict.
- Merging conflict: The joining of two traffic streams.
- Diverge conflict: The separation of two traffic streams, typically the least severe of all conflicts.

Although these descriptions are useful for understanding roundabouts and other intersection types, they can also be both misleading and sometimes inaccurate. The severity of a crash is not a function of the conflict type, but rather a function of the kinetic energy released, relating to the velocity differential between the colliding vehicles. For example, a driver making a right turn onto a high-speed roadway (a "merging conflict") is at risk of a severe crash if struck by highway traffic while turning or shortly thereafter. Conversely, two low-speed traffic streams that cross at a shallow angle may constitute a "crossing conflict," but the low speeds and shallow angle of collision would make severe damage or injury unlikely. Furthermore, the term "merging" can inadvertently and incorrectly imply driver behaviors often not appropriate to roundabout entry, including speed-matching and lane changes, such as occurs at ramp entrances to controlled-access highways.

Therefore, although an understanding of conflict points and their configurations is important to any discussion of roundabout conflict areas, it will be helpful to first repackage and redefine these conflict point types into a more correct and universally applicable format, while keeping with the spirit of conflict points as described in NCHRP 672:

> **Crossing conflict:** A conflict point where two through movements intersect, with two input traffic streams and two output traffic streams.
>
> **Joining conflict:** A conflict point where a through movement terminates at another through movement, with two input traffic streams and one output traffic stream.
>
> **Separating conflict:** A conflict point where a through movement originates by departing from another through movement, with one input traffic stream and two output traffic streams.

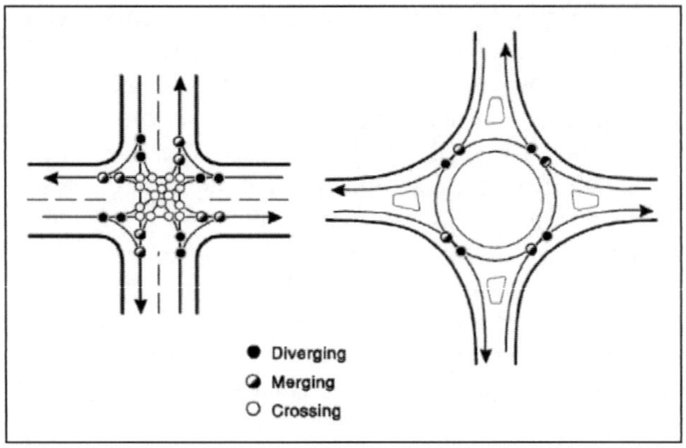

Figure 2. Exhibit 5-2 from NCHRP 672 (2).

Within this framework, traffic operations and vehicle interactions between two traffic streams can be readily explained by ordinary traffic codes, which is particularly useful when designing, operating, and analyzing roundabouts. For the purposes of this discussion, pedestrian conflicts are omitted, as they are generally exterior to the roundabout itself. Additionally, multi-lane one-way carriageways will be treated as one traffic stream for the purpose of counting and categorizing conflict points. Lastly, additional conflict types can and do exist in the driving environment and certain roundabout designs. One such example is a lane reduction conflict, wherein a driver must execute a lateral lane change; however this conflict area spans a considerable distance rather than occurring at a singular location, and therefore is not a "conflict point" for the purposes of this document.

Roundabout Conflict Points

NCHRP 672, Chapter 5, provides a discussion of roundabout conflict points, comparing roundabouts to a standard intersection between two different two-way roadways. NCHRP 672 Exhibit 5-2, entitled "Vehicle Conflict Point Comparison for Intersections with Single-Lane Approaches," reproduced here as Figure 2, provides an example drawing to support the discussion contained in the report. However, whereas NCHRP 672, as a complete document, focuses almost exclusively on crossing roundabout designs, Exhibit 5-2 shows a roundabout with eight conflict points (four merging and four diverging), which is indicative of a concentric roundabout design, for reasons explained later in this section. Notably, this configuration of conflict points is at odds with all roundabout examples contained in the 2009 MUTCD.

Although the MUTCD does not contain a discussion of conflict points, it does provide numerous example figures of roundabout signing and striping, from which traffic streams and movements can be clearly distilled and conflict points and types readily identified. Figure 3 is an example image from Chapter 3 of the 2009 MUTCD showing a typical multi-lane roundabout, with items in red added to the figure for the purpose of this discussion.

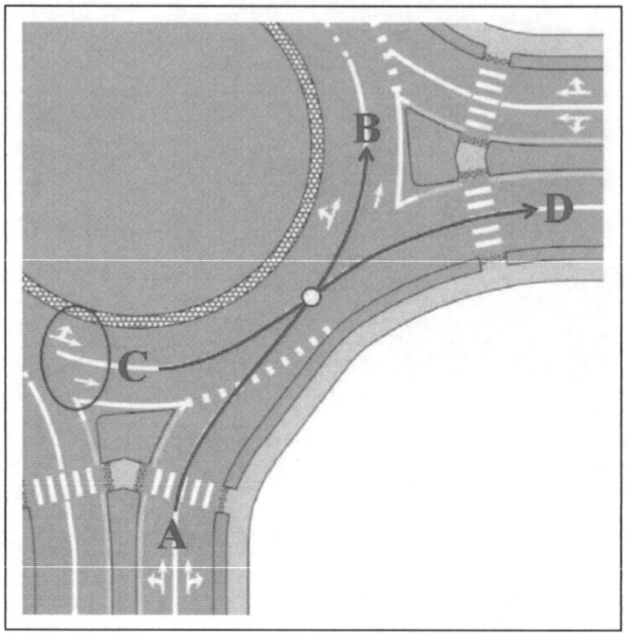

Figure 3. Crossing roundabout conflict configuration. Base image from MUTCD 2009 Edition, Chapter 3 (*1*). Items in red added for illustration purposes.

Note that the pavement arrows shown near Location A in the figure indicate that travel from A to B constitutes a *through* movement. Similarly, the pavement arrows shown near Location C indicate that travel from C to D also constitutes a *through* movement. Accordingly, a movement from C to B constitutes a *left turn* (and therefore a lane departure). Therefore, the through traffic streams from A to B and from C to D form one conflict point, rather than two. A conflict point where two through movements intersect, with two inputs and two outputs as is the case here, is indeed a crossing conflict. Although this example shows a multi-lane roundabout, the same remains true for all other roundabouts, including single-lane roundabouts and mini-roundabouts, illustrated elsewhere in the MUTCD.

Accordingly, a typical crossing roundabout in fact consists of four crossing conflicts, along with eight joining conflicts and eight separating conflicts, for a total of 20 conflict points in all, contrary to what is suggested by NCHRP 672 Figure 5-2. As noted previously, the presence of a crossing conflict, or a greater number of conflicts, is not a reliable indicator of hazard or conflict severity, especially in the case of low speed differentials. However, the configuration of these conflict points is essential to understanding the required driver behaviors in multi-lane situations, because it allows for the conflict points to be modeled as simple interactions between two intersecting traffic streams.

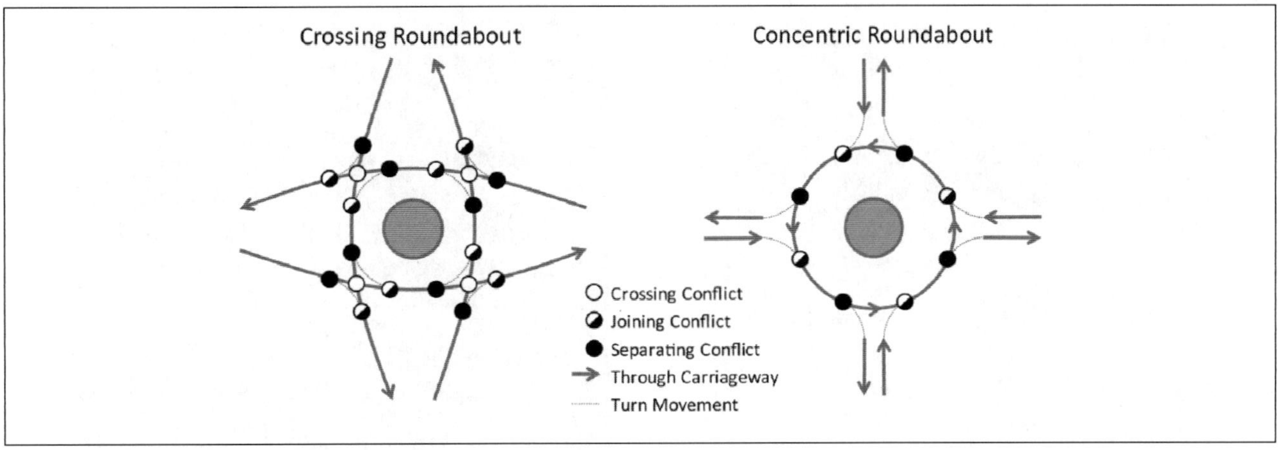

Figure 4. Conflict point diagram comparison.

By contrast, a typical concentric roundabout does indeed consist of eight conflict points, four of which are joining conflicts and four of which are separating conflicts. Each conflict point is formed by an entry or exit traffic stream terminating or originating at a loop carriageway which encircles the central island. A comparison of conflict points for a typical four-legged crossing roundabout versus a four-legged concentric roundabout is shown in Figure 4.

To understand the importance of these distinctions with respect to driver behavior and safety, consider an example, illustrated in Figure 5, wherein each through carriageway consists of two lanes, and a driver wishes to pass through the roundabout from east to west. In the concentric roundabout, every entry point functions as a dual right turn from the stem of a "T" intersection, and therefore a driver approaching via the right-hand lane need only select a gap in the outer (right-hand) lane of the circulatory roadway. In the crossing roundabout, an approaching driver entering via the right-hand lane must yield to both lanes of traffic from the left, and a driver who fails to do so will be at fault for a crash.

Consider now these same example intersections when a driver wishes to exit the intersection heading east, as shown in Figure 6. In the crossing roundabout, both lanes may exit without conflict, but only the lane nearest the center island also has the option to circulate (by turning left, thus departing the through lane). In the concentric roundabout, a driver who wishes to exit from the inner lane of the circular carriageway must first cross the outer ring. The consequences of this weakness can be readily seen in multi-lane concentric roundabouts, where drivers avoid using the left lane in order to avoid potential conflicts or entrapment. "Turbo" roundabouts were developed to address this weakness by physically forcing the outer lane to exit, but still define the exit movement as a *turn*.

Some roundabout intersections, particularly those with more than four approach carriageways, may in fact contain features of both crossing roundabouts and concentric roundabouts. A simple exercise, described in the following section, can be used to help create conflict point framework diagrams for any roundabout, including those with more than four approach carriageways. This exercise can be useful to designers to help to ensure that pavement markings and signing are correct for the configuration, to assist in enforcement of traffic laws, and to enable proper mapping, crash diagramming, and modeling/simulation. Proper framework mapping of roundabout carriageways can also be important for GPS navigation technologies and, in the future, for the safe operation of autonomous vehicles.

The Line and Island Model

The line and island model exercise may be applied to inplace or proposed roundabout designs to determine the operational framework for the entire intersection and for each of the individual conflict areas. The exercise develops a simplified framework of the operating characteristics of a given roundabout using only a series of lines plus a rounded symbol representing the central island. A complete example of this process, using a crossing roundabout with five entry points, is illustrated in Figure 7.

The first step in the line and island model exercise is to identify each of the through carriageways at the roundabout. At existing roundabouts, this may be determined by observing the inplace pavement markings including both lane striping and arrow symbols, or, in the absence of such markings, by considering the prevailing nature of vehicle interactions wherever two through carriageways meet. A through carriageway should always follow the through arrows, and should not cross any lane line pavement markings (except at crossing conflicts where one through movement crosses another). The purpose of this exercise is not to determine the total number of possible conflict points, but rather to clarify how the various through movements interact with each other; therefore, turning movements which do not constitute the

Figure 5. Multi-lane entry operations comparison: (a) crossing roundabout—crossing to enter, and (b) concentric roundabout—turning to enter.

Figure 6. Multi-lane exit operations comparison: (a) crossing roundabout—through movement to exit, and (b) concentric roundabout—turning movement to exit.

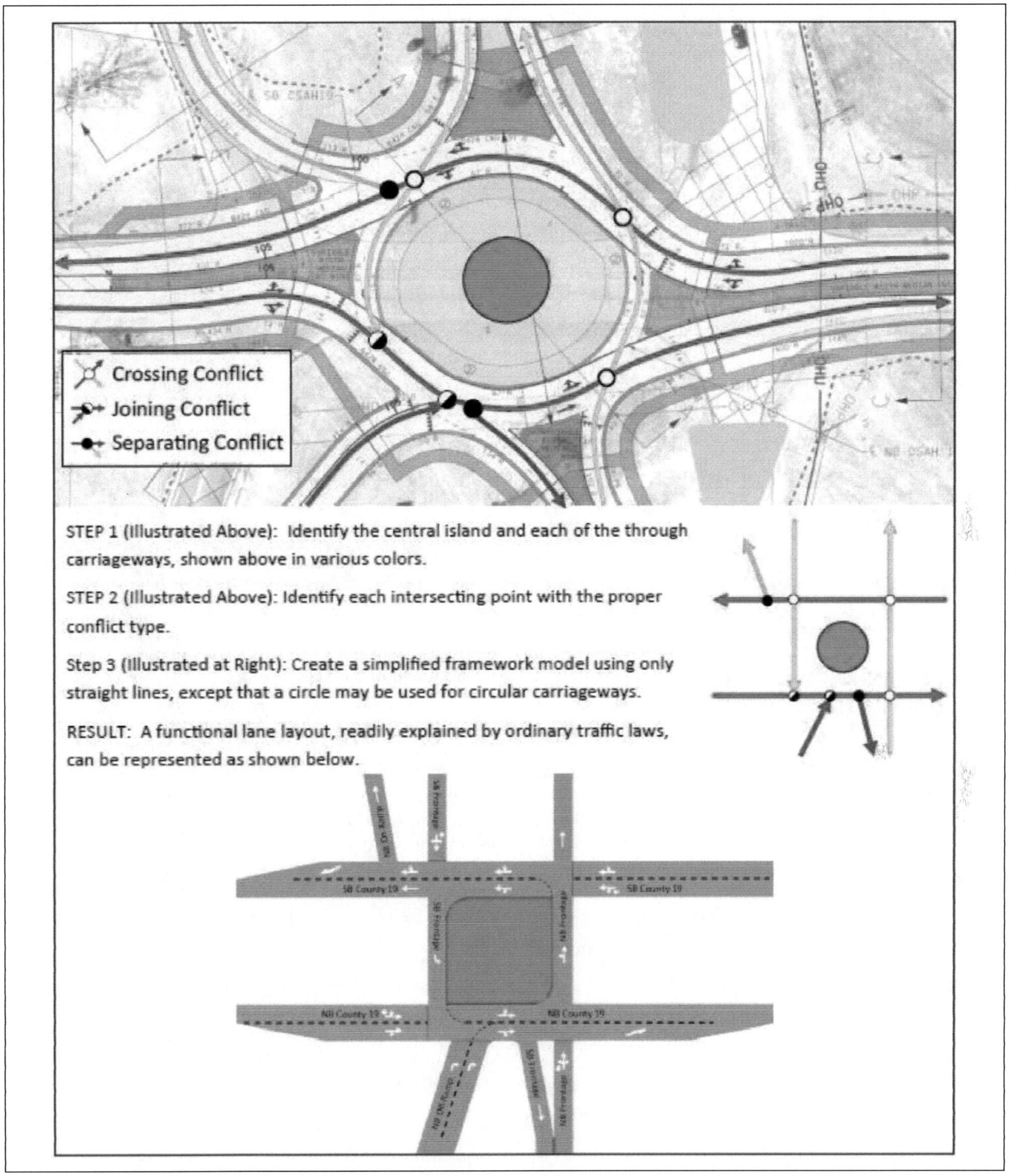

Figure 7. Line and island model exercise example.

origin or termination of through carriageways are not included.

Through carriageways should be continuous to the point at which they either exit the intersection area or terminate at

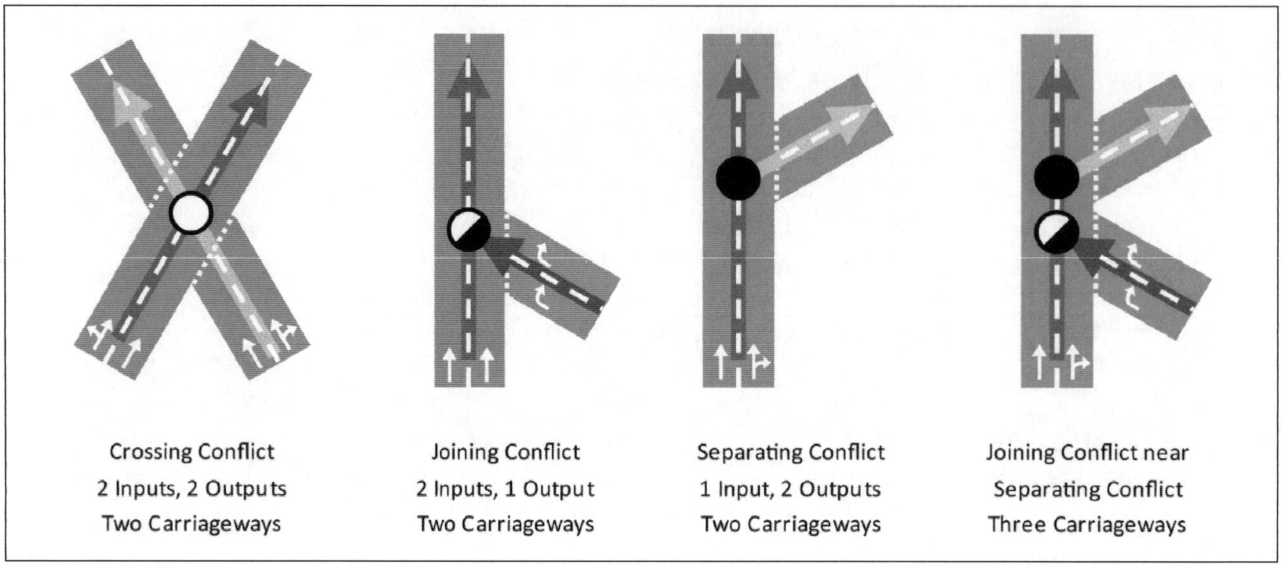

Figure 8. Conflict point carriageway configuration.

a mandatory turn where no through option is available. Carriageway centerlines should follow the centerline of the through traffic stream, and therefore should follow the center of a one-lane carriageway, or the lane line dividing a dual-lane carriageway, or the center of the center lane of a three-lane carriageway.

Once the through carriageways have been identified, identify each location where two carriageways meet with the proper conflict type. As described earlier in this paper:

Crossing conflict: A conflict point where two through movements intersect, with two input traffic streams and two output traffic streams.

Joining conflict: A conflict point where a through movement terminates at another through movement, with two input traffic streams and one output traffic stream.

Separating conflict: A conflict point where a through movement originates by departing from another through movement, with one input traffic stream and two output traffic streams.

Note that at separating conflicts, one of the output carriageways should be the continuation of the input carriageway. Likewise, at joining conflicts, the output traffic stream should be the continuation of one of the input carriageways, though it need not be the input having the right of way. Crossing conflicts only include the intersection of two through carriageways forming an "X" pattern, and do not include closely-spaced joining and separating conflicts forming a "K" pattern (consisting of three through carriageways), as illustrated in Figure 8.

At this point in the exercise, a typical crossing roundabout should be recognizable, having crossing conflicts surrounding the central island. A typical concentric roundabout should likewise be recognizable at this point as having a continuous circular carriageway surrounding the central island, with only joining and separating conflicts surrounding the central island, and no crossing conflicts.

The final step, illustrated as Step 3 in Figure 7, is to simplify the framework diagram to the greatest degree possible by replacing the through carriageways with straightened lines, except for any continuous loop carriageways which necessarily may be represented by a circle, indicative of a concentric roundabout design. This step helps to provide additional clarity about how the roundabout and its conflict areas operate within ordinary traffic regulations (such as the UVC), by dividing its operations into individual turning and crossing movements. Through this step, a crossing roundabout will nearly always form a rectangular shape, with teardrop roundabouts omitting one side of the rectangle. Each individual conflict point, regardless of type, is fully analogous to ordinary interactions between one-way roadways, and each conflict area can have its interactions, and required driver behaviors, readily explained by ordinary traffic laws such as the UVC.

For example, as illustrated in Figure 9, a driver intending to *cross* a carriageway and who is confronted with a yield sign, as shown in (A), must yield to *both* lanes of cross traffic and would not be required to use a turn signal to cross. A driver departing a through carriageway onto a separate departure road as shown in (B) would be required, under typical traffic laws, to take up a position on the proper side of the carriageway in advance of their turn so as not to cross over any adjacent through lanes. A driver approaching a

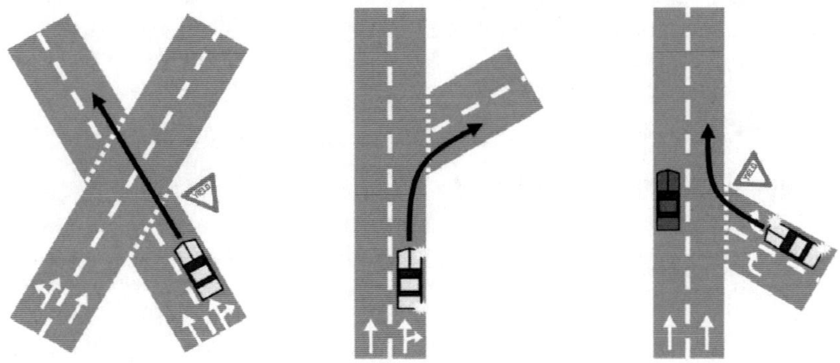

Figure 9. Conflict point operations.

joining conflict on the terminating carriageway as shown in (C) would be able to enter the through carriageway alongside cross traffic, provided that the near lane was clear of traffic. The lane usage and pavement markings on the approach to any given conflict point should be of the same essential configuration as those used in non-roundabout situations.

This exercise is particularly useful for roundabouts which have more than four approach or departure carriageways, as the conflict points surrounding the central island, in the case of crossing roundabouts, will be mixture of crossing conflicts, joining conflicts, and separating conflicts. In the example shown in Figure 7, a crossing roundabout design was utilized, with additional legs added as "T" intersections with the through carriageways, to provide a safer and more effective solution for closely-spaced intersections. The configuration becomes more readily apparent once the carriageways are straightened in the third step of the exercise, thus informing design decisions for pavement markings, signing, and concrete pavement joints.

Summary of Recommended Terminology

Many different definitions of "roundabout," and its many built incarnations, have been developed by various organizations in various regions of the globe, and what constitutes a roundabout in one region or organization may run afoul of the definition of another. Therefore, no universally accepted set of terminology exists.

Based on the operational distinctions presented in this paper, and clarified by the line and island model exercise, it becomes possible to develop a slate of terminology that can be applied to the full universe of "roundabout" designs including all shapes, sizes, operating speeds, and traffic control devices. Using these terms, and understanding these distinctions, can help design and education efforts to fit the operational realities of a particular roundabout. Additionally, these terms and definitions have been developed such that they may apply to all countries, regardless of whether driving occurs on the left-hand or right-hand side of the roadway.

NCHRP Synthesis 264, published in 1998, provides a brief history of circulatory intersections and does include some discussion of the evolution of terminology for such intersections. Notably, "gyratory" is indicated not as a noun but rather as an adjective, used to describe intersections featuring one-way circulation around a central island. The report goes on to note that "roundabout" was officially adopted in Great Britain in 1926 to replace the term "gyratory" (*5*).

One of the most challenging terms to define is also among the most widely used: "Modern roundabout." This term has been widely used to apply to both crossing and concentric designs, despite their operational differences, and assigning the term "modern" to only one of these two varieties would implicitly disparage the other. A review of NCHRP Synthesis 264 suggests that the term refers specifically to circular intersections designed for low-speed, high-capacity throughput at which entering traffic must yield to circulating traffic, as was developed by the United Kingdom's Transportation Research Laboratory (TRL) in the 1960s. Deflection of vehicles through the intersection is also required, though NCHRP 264 does not specify if this deflection must occur on the approaches, or simply by diverting around the central island. The term modern roundabout would thus appear to exclude all roundabouts controlled by traffic signals, though roundabouts with upstream metering signals would not be excluded because the yield-at-entry rule remains in effect at all times.

Maximum throughput speed is left vague, and therefore is excluded from the "modern roundabout" definition provided in the following section.

The following is a slate of universally applicable definitions as recommended by this paper:

Roundabout: A roadway intersection featuring a central island around which traffic progresses in only one direction.

Modern roundabout: A roundabout at which traffic entering the intersection must yield the right of way to conflicting traffic at the entry point.

Concentric roundabout: A junction formed by a series of three-legged ("T") intersections with a continuous loop carriageway encircling a central island.

Crossing roundabout: A junction between two intersecting roadways, in which turns made across opposing traffic are completed by way of the far side of a rounded central island.

Turbo roundabout: A roundabout design which utilizes added interior lanes along with physical channelization to force traffic to exit the intersection.

Mini roundabout: A roundabout with a fully traversable central island.

Rotary/free-flow roundabout: A roundabout design intended to operate without requiring vehicles to stop or yield at entry, which instead relies on speed-matching and lane changes for drivers to enter the intersection area.

Gyratory: An adjective used to describe the operations of any intersection featuring one-way circular flow around a central feature.

Traffic circle: Any circular junction which is *not* a "modern" roundabout, which therefore includes one or more disqualifying features such as high-speed entry, large diameters, signal control, two-way circulating traffic, or where entering traffic has the right of way.

Circulatory roadway: A continuous through carriageway surrounding the central island of a concentric roundabout, intersected at various points by entry and exit carriageways. A crossing roundabout does not possess a true circulatory roadway, as through carriageways exit the intersection area rather than circulating.

Inscribed roadway area: The area exterior to the central island and truck apron of a roundabout but interior to the inscribed intersection area. At crossing roundabouts, this area includes portions of multiple through carriageways, but typically consists of unified geometric design elements, including turn radii and cross slope from the central island curbing.

Crossing conflict: A conflict point where two through movements intersect, with two input traffic streams and two output traffic streams.

Joining conflict: A conflict point where a through movement terminates at another through movement, with two input traffic streams and one output traffic stream.

Separating conflict: A conflict point where a through movement originates by departing from another through movement, with one input traffic stream and two output traffic streams.

Author Note

Aerial images sourced from Google Maps, used as permitted under Fair Use principles as described at https://www.google.com/permissions/geoguidelines.html

Author Contributions

The author confirms contribution to the paper as follows: study conception and design: JG; data collection: JG; analysis and interpretation of results: JG; draft manuscript preparation: JG. The author reviewed the results and approved the final version of the manuscript.

References

1. U.S. Department of Transportation. *Manual on Uniform Traffic Control Devices for Streets and Highways*, 2009 ed. U.S. Department of Transportation, Washington, D.C., 2009.
2. *NCHRP Report 672: Roundabouts, an Informational Guide*, 2nd ed. Transportation Research Board of the National Academies, Washington, D.C., 2010.
3. Millennium Edition of the Uniform Vehicle Code. *National Committee on Uniform Traffic Laws and Ordinances (NCUTLO)*, Alexandria, VA, 2000.
4. Brilon, W. Roundabouts: A State of the Art in Germany. *Presented at the Transportation Research Board's National Roundabout Conference*, Vail, Colorado; May 22–25, 2005. teachamerica.com/RAB14/RAB14papers/RAB14ppr045_Brilon.pdf, Accessed December 14, 2016.
5. Jacquemart, G. *NCHRP Synthesis of Highway Practice 264: Modern Roundabout Practice in the United States*. TRB, National Research Council, Washington, D.C., 1998.

The Standing Committee on Roundabouts (ANB75) peer-reviewed this paper (18-00542).

Article

TRR
JOURNAL OF THE TRANSPORTATION RESEARCH BOARD

Transportation Research Record
2018, Vol. 2672(34) 73–84

DOI: 10.1177/0361198118776159
journals.sagepub.com/home/trr
SAGE

How Roundabout Entry Design Parameters Influence Safety

Jan Novák[1], Jiří Ambros[1], and Jindřich Frič[1]

Abstract

Roundabouts are considered the safest intersection design; however, the safety effect may not be satisfactory at each specific roundabout. This is true especially in countries where roundabout design is a relatively new concept, such as in the Czech Republic. Specifically, most Czech roundabout crashes were found to occur on entries. This motivated the presented study to investigate how entry design parameters influence safety on Czech roundabouts and, if possible, use the findings to update current Czech roundabout design guidelines. To this end, the study comprised three analyses: crash-based safety performance functions, speed analysis, and finally safety performance functions which incorporated speed. All three analyses proved that entry design parameters have a statistically significant influence on safety, in terms of crash frequency, severity and speeds. Given the study objective, this fact should be considered in Czech roundabout design guidelines.

Intersections, where road users may change their directions to get to their destinations, are crucial for the road network operation. However, they also present a discontinuity in the road network and, therefore, a potential hazard, due to several conflict points and traffic complexity (*1–4*). In the United States, more than 50% of fatal and injury crashes take place at or near intersections (*5*). In Europe, every 5th road fatality is due to intersection crashes (*6*). In this context, roundabouts are considered the safest intersection design, as they introduce few conflict points and low speeds, which are associated with reduction in the number and severity of crashes (*7*). Nevertheless, the safety improvements may not produce satisfactory results at each specific roundabout (*8*); in addition, the crash reductions are most pronounced for motor vehicles and less pronounced for pedestrians or bicyclists (*9*). Other roundabout disadvantages include, for example, difficulties for visually impaired users, or potential increase of single-vehicle and fixed-object crashes (*10*).

One of explanations, offered by Montella (*11*), is that there may be safety issues especially in countries where roundabout design is a relatively new concept. The Czech Republic, where roundabouts have emerged since the early 2000s, may be one such example. In fact, Czech roundabout safety performance was found inferior, compared with other European countries (*12*), which may be due to insufficient roundabout experience, both in terms of design practice and driving performance. Similar to other countries (*9*), most Czech roundabout crashes occur on entries: entering–circulating crashes present 58% of all roundabout crashes (*13*), probably due to failure to give way (*14*).

Given these facts, the presented study aimed to investigate how entry design parameters influence safety on Czech roundabouts. Driving behavior on roundabout entries is likely to be influenced by driving path geometry, sight conditions, and so forth. However, the impact of these influences has not been specifically quantified. The objective was to investigate the relationships between entry design parameters and safety, and if possible, use the findings to update current Czech roundabout design guidelines.

Background

International perception of superior roundabout safety performance comes mostly from studies of crash reductions after converting traditional intersections to roundabouts. For example, a study of 23 conversions in the United States (*15*) found 40% reduction of all crashes, 80% of injury crashes, and 90% of fatal crashes. Meta-analysis of 28 non-U.S. studies (*16*) showed 30% to 50% injury crash reduction; fatal crashes were reduced by

[1]CDV – Transport Research Centre, Brno, Czech Republic

Corresponding Author:
Address correspondence to Jan Novák: jan.novak@cdv.cz

50% to 70%. A Czech before–after study (*12*) identified approximately 50% reduction of both total and injury crashes. And recent meta-regression analysis (*17*), based on 44 international studies, concluded that converting intersections to roundabouts is associated with a reduction of fatal and injury crashes of about 65% and 40%, respectively.

However, roundabout implementations may be influenced for example by capacity requirements or spatial constraints; as a result, final performance may not be as safe as expected. When studying roundabout crash types, the highest percentage usually relates to roundabout entries. In a U.S. study, entering–circulating crashes were found to represent about three quarters of all collisions (*18*); above-average numbers of up to 71% were also identified in samples from the United Kingdom (*19*) and 57% in Switzerland (*8*).

In terms of roundabout geometry, scientific literature provides indications of several safety-relevant characteristics. Some of them are as follows (sorted chronologically):

- UK study of 4-leg roundabouts (*19*) reported several safety-related geometric variables, including entry path curvature, entry path radius (the inverse of entry path curvature) and entry width.
- According to German experience (*20*), sufficiently large Δ parameter (defined as the distance between the straightest line from an entry to the opposite exit and the shortest track a driver could take on the circulating roadway) is the precondition to necessary speed reduction. Similar concept was referred to as "lateral displacement" (*21*) or "lateral deflection" (*22*).
- Approach curvature, central island diameter, separation between legs and other factors were found influential in an Australian study (*23*).
- Swiss research (*8*) found correlation between smaller deviation angles and higher crash rates, caused by failing to give way and increased through-speeds.
- International comparison of roundabout design guidelines (*24*) concluded that the main roundabout safety determinant is a combination of entry deflection and entry width.
- A detailed study of roundabout crash contributory factors (*14*) stressed the crucial role of a moderate radius of deflection and a large deviation angle.

In principle, almost all the mentioned parameters are related to deflection, either on entries (entry angle, i.e., the conflict angle between the entering and the circulating traffic) or in the roundabout center (deviation angle, i.e., the amount of trajectory change imposed by the central island). Reducing the vehicle path radius at the entry (i.e., deflecting the vehicle path) decreases the relative speed between entering and circulating vehicles and thus results in lower entering–circulating vehicle crash rates (*7*). These facts have been reflected in several guidelines and standards, for example:

- Spanish standards (*25*) set range of entry angle values 20°– 60° (ideally 30°); the same range is required by UK *Design Manual for Roads and Bridges* (*26*).
- Swiss and Italian standards require a deviation angle above 45° (*27*, *28*).

Regarding safety requirements, Czech guidelines recommend verification of through-speeds; however, they do not present any specific values of entry angle or deviation angle. This study aims to fill this gap by investigating the relationships between entry design parameters and safety, and using the findings to update current Czech roundabout design guidelines.

To this end, the study comprised three analyses, which are described in the following text.

Analyses

Sound research might provide meaningful insight to improve geometric design standards and guide towards the optimal balance between the conflicting design parameters (*28*). To take account of the multi-factor character of the issue, the first step is a multivariate analysis, which considers simultaneously the effects of many factors on the incidence of crashes (*29*). A suitable tool for this task is a mathematical equation representing the number of crashes as a function of the explanatory variables (potential risk factors). These equations are referred to as crash prediction models or safety performance functions (SPFs). In the case of roundabouts, SPFs may be of two kinds (*30*):

- Intersection-level SPFs relate the crash prediction to annual average daily traffic (AADT) and possibly other context variables, such as number of lanes or number of legs.
- Approach-level SPFs relate common types of crashes (e.g., approaching, entering–circulating, or exiting–circulating crashes) to specific AADTs and key geometric parameters (risk factors).

Analysis 1—Safety Performance Functions

Review. Developing SPFs, as a tool of quantitative road safety management, is not a new activity. However, SPFs are usually defined for typical network elements (road segments and intersections). Many countries do not use any specific roundabout SPFs. For example, in the United States, although several roundabout SPFs were developed under NCHRP Project 3-65 (*30*). None of them was introduced in the current edition of *Highway Safety Manual* (HSM) (*31*). The on-going NCHRP Project 17-70 aims to develop roundabout SPFs to be used in the second edition of HSM (*32*).

On the contrary, in New Zealand, with a long tradition both in roundabout constructions and using SPFs, approach-level SPFs are firmly established and implemented in national evaluation guidelines (*33*). Also, Nordic countries (Denmark, Sweden, Finland) are all using specific roundabout SPFs (*34–36*); however only on the intersection level. In the Czech Republic, roundabout SPFs were developed as well (*37–39*); however no Czech approach-level SPF exists.

Data. Given the availability of AADT data on all legs, a sample of 200 typical (unsignalized) roundabouts was selected for analysis. To study individual entries, each roundabout was split into individual leg segments.

Crash data were retrieved from 8 years (2009–2016), for a distance of 100 m (approx 300 ft) from the roundabout center, based on Avelar et al. (*40*). In the Czech Republic, crash reporting is not routinely linked to specific types, such as entering–circulating crashes; therefore, all crash types were used. Regarding their severity, approximately 77% were property-damage-only (PDO) crashes. Distinguishing individual severity levels would thus result in small samples. Therefore, an alternative approach was tested: using national values of crash costs for severity levels (*41*), equivalent of PDO crashes (EPDO) was calculated (value 1 represents one PDO crash, and values 3.68, 32.70, 97.38 represent one slight, severe and fatal injury, respectively). In the analysis, both crash frequency and EPDO were used as a response variable.

As mentioned, AADT from national traffic census was used as a source of traffic volume data. This data however do not provide disaggregation into specific movements, such as entering, circulating, or exiting AADT.

Road and geometric characteristics were collected from online maps. During collection, it was found that conditions were changing during the 8-year period: there were cases in which, for example, a bypass, pedestrian crossing, or even another roundabout leg was added. Therefore, each year was considered individually. AADT values were interpolated between years, according to national traffic forecasting guidelines (*42*). Also, EPDO was calculated separately for each year.

In summary, the sample of 200 roundabouts comprised 781 approaches, separated into 8 annual records. After some data reduction, these resulted in a total of 5,193 data records. Based on literature review, the following road and geometric variables were assigned to all the records:

- Intersection-level characteristics:
  - Location (rural or urban);
  - Roundabout type (single-lane, double-lane, mini, turbo, grade separated);
  - Circular shape (no or yes);
  - Number of legs (3, 4, 5, or 6 legs);
  - Inscribed circle and central island diameter;
  - Standard deviation (SD) of angles between legs (SD is zero for perpendicular legs; the variable shows the difference from the ideal configuration);
  - Number of circulatory lanes (1 or 2);
  - Circulatory lane width; and
  - Truck apron width.
- Approach-level characteristics:
  - Entry angle, Deviation angle, Exit angle (as previously defined);
  - Entry width, Exit width, Bypass width;
  - Close proximity features (binary variables showing presence of public transport stop, parking, accesses or intersections, within 100 m from roundabout);
  - Number of entry lanes and Number of exit lanes;
  - Pedestrian crossing (no or yes);
  - Driving directions (entry, exit, or both);
  - Alignment offset (offset of leg alignment from the radial direction); and
  - Collision distance (distance between entry and following exit).

Exploratory Analysis. The next step was exploratory analysis. During trials, it was found that categorization of some continuous variables improved their relationship to crash frequency. Therefore, selected variables were transformed as follows:

- Entry angle was categorized into 5 intervals: <20°, (20°; 40°), (40°; 60°), (60°; 80°), and >80°.
- Bypass width, Truck apron width, and Alignment offset were transformed into binary absence or presence (no or yes).
- Instead of entry width, variable Entry type was created, using number of entry lanes and number of circulatory lanes:
  - E1 is 1 entry lane and 1 circulatory lane;
  - E2 is 1 entry lane and 2 circulatory lanes; and
  - E3 is 2 entry lanes and 2 circulatory lanes.

Table 1. List of Variables, Transformations and Selection for Modeling

Variable	Transformation	Modeling input
Intersection-level characteristics		
Location		Location
Roundabout type		Roundabout type
Circular shape		Circular shape
# legs		
Inscribed circle diameter		Inscribed circle diameter
Central island diameter		
SD of angles between legs		SD of angles between legs
# circulatory lanes		# circulatory lanes
Circulatory lane width	Entry type categories	
Truck apron width	Apron presence	Apron presence
Approach-level characteristics		
Angles		
Entry angle	Entry angle intervals	Entry angle intervals
Deviation angle		
Exit angle		
Widths		
Entry width	Entry type categories	
Exit width	Entry type categories	
Bypass width	Bypass presence	Bypass presence
Close proximity features		
Public transport stop		Public transport stop
Parking		Parking
Access		Access
Different location		Different location
Close intersection type		Close intersection type
Other		
Number of entry lanes		Number of entry lanes
Number of exit lanes		Number of exit lanes
Pedestrian crossing		Pedestrian crossing
Driving directions		
Alignment offset	Offset presence	
Collision distance		Collision distance

Next, correlations between all pairs of variables were checked. If correlation coefficient exceeded 0.5 threshold, one of the variables was removed. After these initial steps, 19 remaining variables were prepared for modeling. Table 1 summarizes this selection process.

Table 2 summarizes descriptive characteristics of the data.

Modeling. Reported data was used to develop SPFs. Consistent with the literature (*19*), the basic form comprised only exposure (AADT), with further explanatory variables (x_i) being added stepwise, keeping only the ones with achieved statistical significance below 5 %

$$\hat{N} = \exp(\beta_0) \cdot (\text{AADT})^{\beta_1} \cdot \exp\left(\sum_{i=2}^{n}(\beta_i \cdot x_i)\right) \quad (1)$$

where

$\hat{N}$ = expected annual crash frequency or crash severity (EPDO),

AADT = exposure variable (daily traffic volume),

x_i = other explanatory variables (risk factors), and

β_i = regression coefficients, to be estimated in modeling.

Generalized linear modeling procedure in IBM SPSS was applied, using a negative binomial error structure with a logarithmic link function. AADT thus took the form of natural logarithm ln(AADT). The linearized model form is

$$\ln(\widehat{N}) = \beta_0 + \beta_1 \cdot \ln(\text{AADT}) + \sum_{i=2}^{n}(\beta_i \cdot x_i) \quad (2)$$

Results. During modeling, several insignificant variables were discarded. Table 3 reports parameters of the final SPFs (crash frequency SPF and crash severity SPF): regression coefficients β_i and achieved levels of statistical significance (Sig.). Statistical significance was below 5% in most cases, with two exceptions (in bold), which had levels between 5% and 10%.

Table 2. Descriptive Characteristics of Continuous and Categorical Variables

		Min.	Max.	Mean	SD
Continuous variables					
Crash frequency		0	16	0.47	1.14
Crash severity (EPDO)		0	113.42	0.83	3.60
AADT [veh/day]		5	24520	4541.04	2813.22
Collision distance [m]		0	106	16.90	10.87
Inscribed circle diameter [m]		19	139	42.20	21.44
Std. dev. of angles between legs [°]		0	61	15.48	11.60
Categorical variables					
Access				no	yes
				42.1%	57.9%
Apron presence				no	yes
				21.3%	78.7%
Bypass presence				no	yes
				93.0%	7.0%
Circular shape				no	yes
				4.7%	95.3%
Close intersection type		X	T	Y	staggered
		18.2%	60.5%	3.2%	1.8%
		star	roundabout	grade-separated	none
		0.5%	8.9%	4.4%	2.4%
Different location				rural	urban
				21.1%	78.9%
Entry angle	0°;20°>	(20°;40°	(40°;60°>	(60°;80°>	none (exit only)
	14.4%	47.2%	33.6%	3.5%	1.3%
Entry type		none (exit only)	E1	E2	E3
		1.2%	88.1%	5.5%	5.2%
Location				rural	urban
				20.7%	79.3%
Number of circulatory lanes				1	2
				90.5%	9.5%
Number of entry lanes			0 (exit only)	1	2
			1.3%	94.4%	4.3%
Number of exit lanes			0 (entry only)	1	2
			1.0%	95.4%	3.6%
Parking				no	yes
				84.2%	15.8%
Pedestrian crossing				no	yes
				39.1%	60.9%
Public transport stop				no	yes
				85.1%	14.9%
Roundabout type	single-lane	double-lane	mini	turbo	grade-separated
	88.5%	8.4%	2.0%	0.7%	0.4%

Interpretation. Regression coefficients in Table 3 are consistent between Crash frequency SPF and Crash severity (EPDO) SPF. Using the latter enabled identifying also the influence of Location and Pedestrian crossing, which were not statistically significant in the former SPF.

Signs of regression coefficients β_i enable the interpretation of directions of influence of individual variables on response variable:

- Positive relationship means that change of a variable is associated with change of response variable in the same direction. Therefore, increasing variable increases crash frequency or severity, and decrease of variable decreases crash frequency or severity.
- Negative relationship means that change of a variable is associated with change of response variable in the opposite direction. Therefore, increasing variable decreases crash frequency or severity, and decrease of variable increases crash frequency or severity.

Effects of categorical variables are to be interpreted in comparison with the reference category (i.e., the one with zero regression coefficient). For example, missing pedestrian crossing ("no") has a negative coefficient, thus it is

Table 3. Parameter of Crash Frequency SPF and Crash Severity (EPDO) SPF

		Crash frequency SPF		Crash severity (EPDO) SPF	
Variable	Category	β_i	Sig.	β_i	Sig.
(Intercept β_0)		−2.800	*0.000*	−2.940	*0.000*
ln (AADT)		0.583	*0.000*	0.583	*0.000*
Collision distance		−0.005	*0.012*	−0.009	*0.014*
SD of angles between legs		0.005	*0.024*	0.006	*0.028*
Entry angle	<20°	−0.952	*0.002*	−0.946	***0.097***
	(20°; 40°>	−1.183	*0.000*	−1.239	*0.028*
	(40°; 60°>	−1.169	*0.000*	−1.212	*0.031*
	(60°; 80°>	−1.122	*0.000*	−1.178	*0.041*
	>80°	0		0	
Apron	no	0.560	*0.000*	0.703	*0.000*
	yes	0		0	
Bypass	no	−0.498	*0.000*	−0.521	*0.000*
	yes	0		0	
Entry type	E1	−1.813	*0.000*	−1.996	*0.000*
	E2	−1.123	*0.000*	−1.277	*0.000*
	E3	0		0	
Location	rural			0.188	*0.017*
	urban			0	
Pedestrian crossing	no			−0.133	***0.068***
	yes			0	

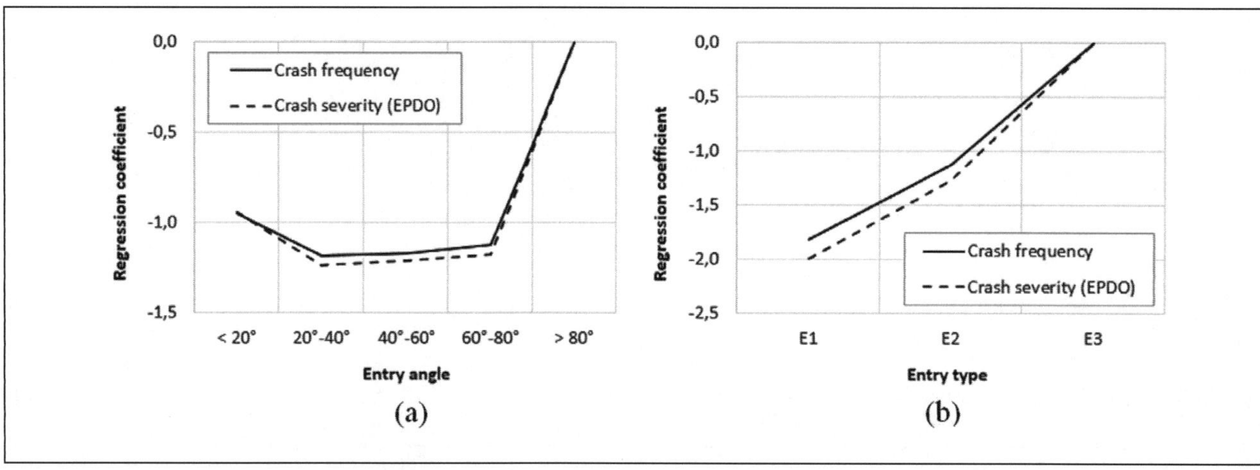

Figure 1. Regression coefficients for categories of (*a*) entry angle and (*b*) entry type.

associated with lower crash frequency, compared to the reference category with pedestrian crossing ("yes").

According to Table 3, the variables had the following directions of influence:

- AADT has a positive influence as expected. In addition, coefficient 0.583 is close to values 0.576 from previous Czech SPF (*39*) or 0.58 from New Zealand SPF (*33*, Table 15).
- Collision distance has a negative association: the longer the distance, the more space for potential evasive maneuvers and lower crash frequency or severity. This was confirmed also by Maycock & Hall (*19*) or Arndt (*23*).
- SD of angles between legs has a positive relationship (the more dispersion, the more complex the environment and the higher the crash frequency or severity), consistent with Czech design standards. Also, Jensen (*43*) noted that "as the difference between the smallest and largest angle between arms in a roundabout becomes smaller, the level of safety becomes better."
- Regression coefficients for entry angle are displayed in Figure 1*a*. The curve has minimum values between 20° and 80°. This is relatively consistent

with Spanish and British standards, which recommend interval of entry angles 20° to 60° (*25*, *26*).

- Figure 1*b* provides analogical graph for entry types. The trend is rising from E1 (single entry and circulatory lane) to E3 (two entry and circulatory lanes). Positive association between entry width (number of lanes) and crash frequency was identified in several studies (*9*, *19*, *37*, *44*, *45*).
- Truck apron has a protective influence: its presence is associated with lower crash frequency or severity. This was also confirmed by Šenk & Ambros (*37*).
- On the contrary, bypass presence is associated with higher crash frequency or severity. This is probably due to adding another conflict point, which increases crash risk and severity, as noted by Robinson et al. (*9*) or Daniels et al. (*46*).
- Location in rural areas seems to increase crash severity, compared to urban roundabouts; the same holds for presence of pedestrian crossing. Both variables are likely to be associated with speed and vulnerability, as also reported by Turner et al. (*44*) or Šenk & Ambros (*37*).

Location variable is likely to be a proxy for speed, which was not considered in the SPF. In this regard, it would be interesting to measure speed and use it in modeling. Therefore, an alternative approach was used for the second analysis step: speeds were measured on a sub-sample of roundabouts and used to explore its relationship to roundabout geometry.

Analysis 2—Speed

Review. Traditionally, speed measurement has been performed using stationary devices. For example, Rodegerdts et al. (*30*) used a speed gun to measure free-flow speeds in specific points of through-pass trajectory. A similar approach was used also by Spacek (*8*), Turner et al. (*44*), and Kim & Choi (*47*). In addition to speed guns, some measured speeds of isolated vehicles also by video camera (*48*, *49*). However, with these methods speeds are collected only in limited spots. Therefore, alternative methods used video detection or image processing to obtain complete speed profiles (*50–52*). Other studies used speed data collected by car-following (*21*, *53*) or by instrumented vehicles (*54*).

Data. In this study, the CDV – Transport Research Centre instrumented vehicle was used. It is a common van, equipped by data collection sensors (GPS, accelerometers, gyroscopes, video camera). GPS location was operating at 10 Hz frequency (10 records per second), with horizontal precision of 1 m. It was used to obtain data points, representing vehicle path and continuous speed. Data were collected on a sub-sample of 11 roundabouts, where two drivers passed several times through each leg, totaling 92 through-drives from one leg to the opposite one. Driving was conducted in free-flow conditions (morning and evening off-peak periods).

Collected data were processed to obtain relevant radii and angles, distinguish tangent and curve segments and estimate respective speeds. The obtained parameters were

- Radii on entry, center and exit;
- Entry angle, deviation angle, exit angle; and
- Speeds in seven profiles (A to G):
  - A, on approach leg (50 m upstream of entry);
  - 5 profiles at a roundabout (B, C, D, E, F); and
  - G, on exit leg (50 m downstream of exit).

The parameters, together with an example of speed profiles, are displayed in Figure 2.

Results. To assess the influence of radii and angles on speed, Spearman correlation coefficients were calculated. Unfortunately, only some exceeded 0.5 threshold, which is perceived as a threshold of moderate correlation (*55*). These were correlations between

- speed 50 m upstream of entry and entry angle;
- speed 50 m upstream of entry and entry radius; and
- speed in center and central radius.

The first two associations relate to roundabout entries, which indicates their relationship to safety. In the first analysis, link between geometry and safety was established through crash frequency or severity SPFs; the second analysis showed the link between geometry and speed. One may thus anticipate a causal chain: geometry – speed – safety. To confirm existence of this chain, the third analysis attempted to develop SPF, which contain both geometry and speed variables.

Analysis 3—Safety Performance Functions with Speed

Review. Several international studies employed speed variables in their SPFs (e.g., *23*, *44*, *56*, *57*). On the other hand, some of the studies did not succeed, for example NCHRP Project 3-65 (*30*), in which the estimated model was deemed inadequate on the basis of weak effects of speed variables. One of the problems may be the use of different speed definitions, such as speed limits, measured speeds, predicted or simulated speeds, and so forth. Other authors attempted estimating speed from geometric parameters, and then using it to model crash frequencies (*49*, *58*, *59*). These studies differed in using

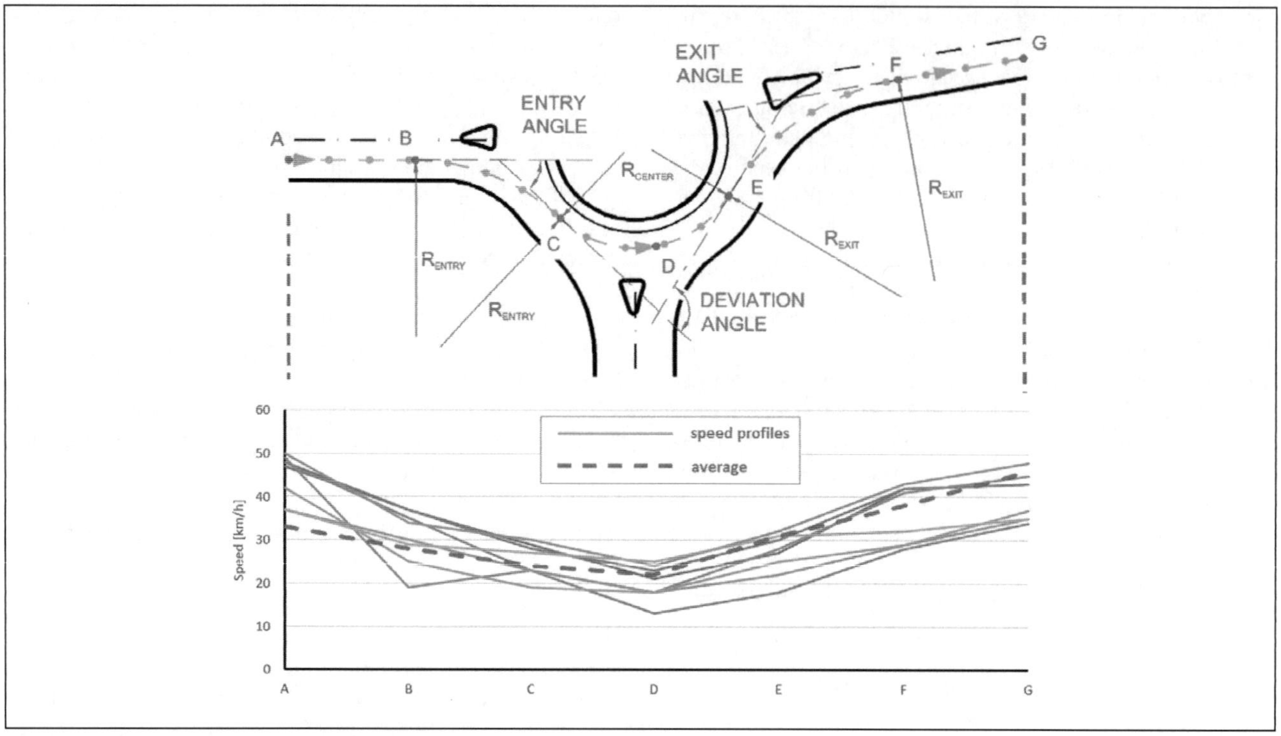

Figure 2. Through-pass trajectory parameters and example of speed profiles.

Table 4. Descriptive Characteristics of Crash, Geometry and Speed Data

Variable	Unit	Min.	Max.	Mean	SD
Crash frequency	[-]	1	25	9.26	6.60
EPDO	[-]	1	59.22	15.69	14.50
Hourly traffic flow	veh/h	64	1600	548.93	488.00
Entry angle	[°]	2	75	34.37	15.93
Deviation angle	[°]	40	191	79.56	32.04
Exit angle	[°]	8	52	31.48	11.73
Entry radius	[m]	18	113	36.59	18.39
Central radius	[m]	12	28	19.67	3.81
Exit radius	[m]	25	92	38.04	13.80
Speed A	[km/h]	31	54	42.33	6.34
Speed B	[km/h]	23	40	31.63	4.13
Speed C	[km/h]	21	35	26.93	3.35
Speed D	[km/h]	18	30	24.44	2.94
Speed E	[km/h]	23	40	31.04	3.54
Speed F	[km/h]	28	50	38.00	4.92
Speed G	[km/h]	30	64	44.41	8.71

measured or predicted speeds, and in the function form of speed (either power or exponential form). It is thus evident that although developing speed-based SPFs is not a new task, it is still not a formalized approach.

Data. This analysis was based on previously described data collected at 11 roundabouts. The following variables were used:

- Crash frequency or crash severity as a response variable, collected for each through-pass;
- Angles and radii as explanatory variables; and
- Speeds in profiles A to G, as additional explanatory variables.

Instead of AADT, hourly traffic flow in respective trajectories was counted on the site. Descriptive characteristics of data are reported in Table 4.

Table 5. SPF Parameters

Variable	β_i	Sig.
(Intercept β_0)	5.211	*0.022*
ln (hourly traffic flow)	−0.727	*0.007*
Entry angle	−0.052	*0.001*
Speed 50 m upstream of entry	0.066	*0.031*

Results. The model form described in Equation 1 was used (with hourly traffic flow instead of AADT). Apart from traffic flow, the SPF introduced entry angle, and speed 50 m upstream of entry. Parameters of the developed SPF (regression coefficients β_i and achieved levels of statistical significance, all below 5%) are reported in Table 5.

Typically, AADT has a positive relationship to crash frequency; on the contrary, negative coefficient of traffic flow was found in all three SPFs. Although not expected, several studies also had such finding:

- Satterthwaite (*60*) found that rear-end crash rates increased with traffic flow, while single-vehicle rates decreased.
- Hiselius (*61*) identified similar confusion, when mixing volumes of cars and trucks.
- Analysis by Christoforou et al. (*62*) concluded that low volume (free-flow) conditions may lead to higher speeds, which in turn cause unexpected direction of traffic flow influence on safety. The unexpected sign of regression coefficient may be thus due to using aggregated input data.

Signs of other regression coefficients are as expected from literature:

- Entry angle has a protective effect: the higher the angle, the lower the speed and risk.
- Speed on approach has a positive relationship to crashes: increasing speed is associated to increasing crash frequency.

Discussion and Conclusions

The study comprised three analyses: crash-based safety performance functions (SPFs), speed analysis, and finally SPFs which incorporated speed. The analyses were different in terms of their data needs: whereas crash-based SPFs required crashes, traffic volumes and geometry (i.e., data that are usually available network-wide), the second and third approach required speed data, that had to be especially collected. This is the reason samples were different (200 roundabouts in Analysis 1 vs. 11 roundabouts in Analyses 2 and 3). On the other hand, using speed in analyses adds an extra value, since it may be considered a proxy for driving behavior, which is likely to enhance the quality of developed models.

With the aim of investigating the influence of roundabout entry design parameters on safety, the analyses concluded as follows:

- Analysis 1 developed approach-level SPFs. Consistent with the literature, several variables were significantly related to crashes, including entry angle and entry type. Using crash severity (EPDO) in addition to crash frequency also revealed significant influence of location (rural or urban) and presence of pedestrian crossing on roundabout leg.
- In analysis 2, speeds were measured on a subsample of roundabouts and used to explore its relationship to roundabout geometry. Correlation was found between approach speeds (50 m upstream of entry) and entry angle and entry radius.
- Analysis 3 successfully incorporated approach speed into SPF and confirmed that increasing speed is associated to increasing crash frequency; again, it was demonstrated that entry angle has a protective effect.

Thus, it was proved that entry design parameters have a statistically significant influence on safety, in terms of crash frequency, severity and speeds. The SPF suggested suitable range of entry angle values, which should be considered in Czech roundabout design guidelines TP 135 (*63*). Currently they only recommend a deviation of entry trajectory from entry axis, which may not be sufficient to reach the necessary speed reduction of vehicles entering the circulatory lane. The analysis also identified the influence of entry radius; the same Czech guidelines mention verification of fastest path radii, but without mentioning any specific values of entry or radii.

At the same time, the authors are aware of several limitations:

- Input crash data is likely to be underreported. In addition, routine crash reporting does not allow distinguishing of individual crash types.
- Routinely collected AADT data, which does not consider individual movements, are also not ideal.
- Through-pass trajectory identification was based on circular elements only, without consideration of transition curves.
- Speed data was collected only by one instrumented vehicle and two drivers, which may not provide representative speed behavior data. Sample of 11 roundabouts, where speeds were measured, was also small.

Further activities should aim to expand sample size and consider the mentioned limitations of crash, AADT and speed data. Nevertheless, the presented results indicate the existence of a causal chain: geometry – speed – safety. The study found that

1. Both crash frequency and severity is influenced by roundabout entry geometry;
2. At the same time, the geometry influences driving speeds; and
3. Safety performance is thus dictated by both geometry and speeds.

In future, these points may be further elaborated to develop a "two-stage" tool for roundabout safety analysis (*49*). In its first stage, speed may be modeled based on geometry and used in the second stage as a surrogate safety measure (instead of crash-based measures) to assess safety, even during the planning phase.

Acknowledgments

The authors appreciate the help of their colleagues Lucie Vyskočilová and Ondřej Gogolín with data collection and processing. The study was conducted with the financial support of the Czech Ministry of Education, Youth and Sports under the National Sustainability Programme I project of Transport R&D Centre, using the research infrastructure from the Operation Programme Research and Development for Innovations.

Author Contributions

The authors confirm contribution to the paper as follows: study conception and design: Jan Novák, Jiří Ambros; data collection: Jan Novák; analysis and interpretation of results: Jan Novák, Jiří Ambros, Jindřich Frič; draft manuscript preparation: Jan Novák, Jiří Ambros, Jindřich Frič. All authors reviewed the results and approved the final version of the manuscript.

References

1. Ogden, K. W. *Safer Roads: A Guide to Road Safety Engineering*. Ashgate, Aldershot, 1997.
2. *Road Safety Manual: Recommendations from the World Road Association (PIARC)*. Route2market, Harrogate, 2003.
3. Elvik, R., A. Høye, T. Vaa, and M. Sørensen. *The Handbook of Road Safety Measures*, 2nd ed. Emerald, Bingley, 2009.
4. Tiwari, G., and D. Mohan, eds. *Transport Planning & Traffic Safety: Making Cities, Roads, & Vehicles Safer*. CRC Press, Boca Raton, 2016.
5. Federal Highway Administration. *Intersection Safety*. Federal Highway Administration, https://www.fhwa.dot.gov/research/topics/safety/intersections/. Accessed July 26, 2017.
6. European Commission. *Basic Traffic Safety Facts 2015*. European Commission, https://ec.europa.eu/transport/road_safety/specialist/statistics_en. Accessed July 26, 2017.
7. Rodegerdts, L., J. Bansen, C. Tiesler, J. Knudsen, E. Myers, M. Johnson et al. *NCHRP Report 672: Roundabouts: An Informational Guide*. Transportation Research Board of the National Academies, Washington, D.C., 2010.
8. Spacek, P. Basis of the Swiss Design Standard for Roundabouts. *Transportation Research Record: Journal of the Transportation Research Board*, 2004. 1881: 27–35.
9. Robinson, B. W., ed. *Roundabouts: An Informational Guide*. Report FHWA-RD-00-067. Federal Highway Administration, U.S. Department of Transportation, Washington, D.C., 2000.
10. Isebrands, H. N., ed. *Enhancing Intersection Safety Through Roundabouts: An ITE Informational Report*. ITE, Washington, D.C., 2008.
11. Montella, A. Roundabout In-Service Safety Reviews: Safety Assessment Procedure. *Transportation Research Record: Journal of the Transportation Research Board*, 2007. 2019: 40–50.
12. Ambros, J., R. Turek, and Z. Janoška. Safety Evaluation of Czech Roundabouts. *Advances in Transportation Studies*, Vol. 40, 2016, pp. 111–122. http://dx.doi.org/10.4399/97888548970079
13. Slabý, P., J. Kocourek, and D. Kočárková. *Analýza vlivu vybraných stavebních opatření. Part IV of BESIDIDO Project Final Report*. Czech Technical University in Prague, Prague, 2005.
14. Montella, A. Identifying Crash Contributory Factors at Urban Roundabouts and Using Association Rules to Explore Their Relationships to Different Crash Types. *Accident Analysis & Prevention*, Vol. 43, 2011, pp.1451–1463. http://dx.doi.org/10.1016/j.aap.2011.02.023
15. Persaud, B. N., R. A. Retting, P. E. Garder, and D. Lord. Safety Effect of Roundabout Conversions in the United States: Empirical Bayes Observational Before-after Study. *Transportation Research Record: Journal of the Transportation Research Board*, 2001. 1751: 1–8.
16. Elvik, R. Effects on Road Safety of Converting Intersections to Roundabouts: Review of Evidence from Non-US Studies. *Transportation Research Record: Journal of the Transportation Research Board*, 2003. 1847: 1–10.
17. Elvik, R. Road Safety Effects of Roundabouts: A Meta-analysis. *Accident Analysis & Prevention*, Vol. 99, 2017, pp. 364–371. http://dx.doi.org/10.1016/j.aap.2016.12.018
18. Mandavilli, S., A. McCartt, and R. Retting. Crash Patterns and Potential Engineering Countermeasures at Maryland Roundabouts. *Traffic Injury Prevention*, Vol. 10, 2009, pp. 44–50. http://dx.doi.org/10.1080/15389580802485938
19. Maycock, G., and R. D. Hall. *Accidents at 4-Arm Roundabouts*. Laboratory Report 1120. Transport and Road Research Laboratory, Crowthorne, 1984.
20. Brilon, W., and M. Vandehey. *Roundabouts – The State of the Art in Germany*. ITE Journal, 1998, pp. 48–54.
21. Hydén, C., and A. Várhelyi. The Effects on Safety, Time Consumption and Environment of Large Scale Use of Roundabouts in an Urban Area: A Case Study. *Accident*

Analysis & Prevention, Vol. 32, 2000, pp. 11–23. http://dx.doi.org/10.1016/S0001-4575(99)00044-5

22. Overkamp, D. P., and W. van der Wijk. *Roundabouts – Application and Design: A Practical Manual.* Ministry of Transport, Public Works and Water Management, The Hague, 2009.
23. Arndt, O. K. *Relationship Between Unsignalised Intersection Geometry and Accident Rates*. PhD thesis. Queensland University of Technology, Brisbane, 2004.
24. Kennedy, J. *International Comparison of Roundabout Design Guidelines*. Published Project Report PPR206. Transport Research Laboratory, Crowthorne, 2007.
25. *Recomendaciones sobre glorietas*. Dirección General de Carreteras, Madrid, 1999.
26. *Design Manual for Roads and Bridges, TD 16/07*. http://www.standardsforhighways.co.uk/ha/standards/dmrb/vol6/section2/td1607.pdf. Accessed July 26, 2017.
27. *Knoten mit Kreisverkehr / Carrefours giratoires*. Swiss Standard SN 640 263. Association of Swiss Road and Traffic Engineers, Zurich, 1999.
28. Montella, A., S. Turner, S. Chiaradonna, and D. Aldridge. International Overview of Roundabout Design Practices and Insights for Improvement of the Italian Standard. *Canadian Journal of Civil Engineering*, Vol. 40, 2013, pp. 1215–1226. http://dx.doi.org/10.1139/cjce-2013-0123
29. Arndt, O., and R. Troutbeck. Techniques for Analysing the Effect of Road Geometry on Accident Rates Using Multifactor Studies. *Proc., 22nd ARRB Conference*, Canberra, 2006.
30. Rodegerdts, L., M. Blogg, E. Wemple, E. Myers, M. Kyte, M. Dixon, G. List, A. Flannery, R. Troutbeck, W. Brilon, N. Wu, B. Persaud, C. Lyon, D. Harkey, and D. Carter. *NCHRP Report 572: Applying Roundabouts in the United States*. Transportation Research Board of the National Academies, Washington, D.C., 2007.
31. *Highway Safety Manual*, 1st ed. AASHTO, Washington, D.C., 2010.
32. Foster, N. Development of Roundabout Crash Prediction Models and Method. NCHRP Project 17-70. *Proc., 5th International Conference on Roundabouts*, Green Bay, 2017.
33. *Crash Estimation Compendium (New Zealand Crash Risk Factors Guideline)*. NZ Transport Agency, Wellington, 2016.
34. Jensen, S. U. Uheldsmodeller for rundkørsler. *Trafik & Veje*, 2013, pp. 54–55.
35. *Bygg om eller bygg nytt, Kapitel 6 Trafiksäkerhet*. Trafikverket, Borlänge, 2016.
36. Peltola, H., R. Rajamäki, and J. Luoma. A Tool for Safety Evaluations of Road Improvements. *Accident Accident Analysis & Prevention*, Vol. 60, 2013, pp. 277–288. http://dx.doi.org/10.1016/j.aap.2013.04.008
37. Šenk, P., and J. Ambros. Estimation of Accident Frequency at Newly-built Roundabouts in the Czech Republic. *Transactions on Transport Sciences*, Vol. 4, 2011, pp. 199–206. http://dx.doi.org/10.2478/v10158-011-0018-4
38. Novák, J., and J. Ambros. Rozšíření predikčního modelu nehodovosti na okružních křižovatkách. *Proc., 20 Silniční konference*, Plzeň, 2012.
39. Ambros, J., and P. Slabý. Comparison of Roundabout Accident Prediction Models: Challenges of Data Collection, Analysis and Interpretation. *Proc., 20th Anniversary of the Faculty of Transportation Sciences*, Prague, 2013.
40. Avelar, R. E., K. K. Dixon, and P. Escobar. Evaluation of Signalized-Intersection Crash Screening Methods Based on Distance from Intersection. *Transportation Research Record: Journal of the Transportation Research Board*, 2015. 2514: 177–186.
41. Vyskočilová, A., O. Gogolín, and O. Valach. New Approach to Evaluation of Socio-economic Losses Caused by Traffic Accidents. Presented at 93rd Annual Meeting of the Transportation Research Board, Washington, D.C., 2014.
42. Bartoš, L., A. Richtr, J. Martolos, and M. Hála. *Prognóza intenzit automobilové dopravy, II. vydání*. Technical Guidelines 225. EDIP s.r.o., Plzeň, 2012.
43. Jensen, S. U. Safety Effects of Height of Central Islands, Sight Distances, Markings and Signage at Single-lane Roundabouts. *Proc., 5th International Symposium on Highway Geometric Design*, Vancouver, 2015.
44. Turner, S. A., A. P. Roozenburg, and A. W. Smith. *Roundabout Crash Prediction Models*. Research Report 386. NZ Transport Agency, Wellington, 2009.
45. Ambros, J., J. Novák, A. Borsos, E. Hóz, M. Kieć, Š. Machciník, and R. Ondrejka. Central European Comparative Study of Traffic Safety on Roundabouts. *Transportation Research Procedia*, Vol. 14, 2016, pp. 4200–4208. http://dx.doi.org/10.1016/j.trpro.2016.05.391
46. Daniels, S., T. Brijs, E. Nuyts, and G. Wets. Extended prediction models for crashes at roundabouts. *Safety Science*, Vol. 49, 2011, pp. 198–207. http://dx.doi.org/10.1016/j.ssci.2010.07.016
47. Kim, S., and J. Choi. Safety Analysis of Roundabout Designs based on Geometric and Speed Characteristics. *KSCE Journal of Civil Engineering*, Vol. 17, 2013, pp. 1446–1454. http://dx.doi.org/10.1007/s12205-013-0177-4
48. Bassani, M., and E. Sacchi. Experimental Investigation into Speed Performance and Consistency of Urban Roundabouts: An Italian Case Study. *Proc., 3rd International Conference on Roundabouts*, Carmel, 2011.
49. Chen, Y., B. Persaud, E. Sacchi, and M. Bassani. Investigation of Models for Relating Roundabout Safety to Predicted Speed. *Accident Analysis & Prevention*, Vol. 50, 2013, pp. 196–203. http://dx.doi.org/10.1016/j.aap.2012.04.011
50. Guido, G., F. Saccomanno, A. Vitale, V. Astarita, and D. Festa. Comparing Safety Performance Measures Obtained from Video Capture Data. *Journal of Transportation Engineering*, Vol. 137, 2011, pp. 481–491. http://dx.doi.org/10.1061/(ASCE)TE.1943-5436.0000230
51. Mussone, L., M. Matteucci, M. Bassani, and D. Rizzi. An Innovative Method for the Analysis of Vehicle Movements in Roundabouts Based on Image Processing. *Journal of Advanced Transportation*, Vol. 47, 2013, pp. 581–594. http://dx.doi.org/10.1002/atr.184
52. Sadeq, H., and T. Sayed. Automated Roundabout Safety Analysis: Diagnosis and Remedy of Safety Problems. *Journal of Transportation Engineering*, Vol. 142, 2016, 04016062. http://dx.doi.org/10.1061/(ASCE)TE.1943-5436.0000887

53. Gaca, S., and M. Kieć. Speed Management for Local and Regional Rural Roads. *Transportation Research Procedia*, Vol. 14, 2016, pp. 4170–4179. http://dx.doi.org/10.1016/j.trpro.2016.05.388
54. Bastos Silva, A., S. Santos, L. Vasconcelos, Á. Seco, and J. P. Silva. Driver Behavior Characterization in Roundabout Crossings. *Transportation Research Procedia*, Vol. 3, 2014, pp. 80–89. http://dx.doi.org/10.1016/j.trpro.2014.10.093
55. Hinkle, D. E., W. Wiersma, and S. G. Jurs. *Applied Statistics for the Behavioral Sciences*, 5th ed. Cengage Learning, Boston, 2003.
56. Brüde, U., and J. Larsson. What Roundabout Design Provides the highest possible safety? *Nordic Road & Transport Research*, No. 2, 2000, pp. 17–21.
57. Farag, S. G., and I. H. Hashim. Safety Performance Appraisal at Roundabouts: Case Study of Salalah City in Oman. *Journal of Transportation Safety & Security*, Vol. 9, 2017, pp. 67–82. http://dx.doi.org/10.1080/19439962.2016.1199623
58. Chen, Y., B. Persaud, and C. Lyon. Effect of Speed on Roundabout Safety Performance – Implications for Use of Speed as a Surrogate Measure. Presented at 90th Annual Meeting of the Transportation Research Board, Washington, D.C., 2011.
59. Lyon, C., S. Chan, and B. Persaud. A Model for Average Speed Estimation and Crash Prediction Using Vehicle Path Data. *Proc., 4th International Conference on Roundabouts*, Seattle, 2014.
60. Satterthwaite, S. P. *A Survey of Research into Relationships Between Traffic Accidents and Traffic Volumes*. Supplementary Report 692. Transport and Road Research Laboratory, Crowthorne, 1981.
61. Hiselius, L. W. Estimating the Relationship Between Accident Frequency and Homogeneous and Inhomogeneous Traffic Flows. *Accident Analysis & Prevention*, Vol. 36, 2004, pp. 985–992. http://dx.doi.org/10.1016/j.aap.2003.11.002
62. Christoforou, Z., S. Cohen, and M. G. Karlaftis. Integrating Real-Time Traffic Data in Road Safety Analysis. *Procedia – Social and Behavioral Sciences*, Vol. 48, 2012, pp. 2454–2463. http://dx.doi.org/10.1016/j.sbspro.2012.06.1216
63. Smělý, M., M. Radimský, and M. Patočka. *Projektování okružních křižovatek na silnicích a místních komunikacích, 3. vydání*. Technical Guidelines 135. Brno University of Technology, Brno, 2017.

The Standing Committee on Roundabouts (ANB75) peer-reviewed this paper (18-01061).

TRR
JOURNAL OF THE TRANSPORTATION RESEARCH BOARD

Article

Transportation Research Record
2018, Vol. 2672(34) 85–95

DOI: 10.1177/0361198118795011
journals.sagepub.com/home/trr
SAGE

Characteristics and Profiles of Moped Crashes in Urban Areas: An In-Depth Study

Tim De Ceunynck[1], Freya Slootmans[1], and Stijn Daniels[1]

Abstract
In this in-depth study, 167 severe injury crashes involving a moped in urban areas in Belgium in 2013 are analyzed. The study provides an overview of the general crash characteristics of and contributing factors to these moped crashes, and identifies typical crash profiles and their characteristics. The representation of class A mopeds (maximum speed 25 km/h [16 mph]; no driving license needed) and class B mopeds (maximum speed 45 km/h [28 mph]; driving license required) in these crashes is approximately fifty-fifty, which seems to imply an over-representation of mopeds class A. The involved moped riders have an average age of 33 and three-quarters are men. Human factors are by far the largest category of contributing factors in moped crashes, factors related to infrastructure and environment play a moderate role, and vehicle-related factors play only a minor role. The most important human factors are psychological factors and risk assessment errors, the main non-human contributing factors are sight obstructions owing to infrastructure or other vehicles. The following crash profiles, grouping similar types of crashes, were identified: crashes between a vehicle turning off and a moped going straight through (18%); crashes caused by the moped rider's risky behavior (17%); crashes between a moped and another vulnerable road user (13%); crashes resulting from entering or exiting destinations alongside the road (12%); crashes at intersections (other than vehicles turning off) (11%); single-vehicle crashes with loss of control (9%); crashes caused by an error during overtaking (8%); rear-end crashes (3%); and other (9%).

The use of powered two-wheelers (PTW) has increased substantially over the years in many developed countries (*1*). The term PTW covers a wide range of vehicles, but the vast majority of PTW can be classified in two main categories: mopeds, which have a small (max 50cc) engine and restricted top speed; and motorcycles (*2*).

PTW are highly over-represented in crash statistics compared with their traffic exposure, constituting 18% of all traffic fatalities in the EU (*3*). In Belgium, PTW only represent 1.2% of the annual vehicle kilometers, but they represent 13.9% of all crash injuries (mopeds 7.6%; motorcycles 6.3%) (*4*). The risk per kilometer for PTW riders of getting involved in a severe or fatal crash is the highest of all transport modes in Belgium, and more than 50 times the risk of an average car driver (*5*). Internationally, studies have shown that the death rate of PTW riders is about 30 times higher than the death rate of car occupants, and both the fatality rate and the injury rate of PTW riders are worse than for car drivers no matter the denominator (registered vehicles, licensed riders, distance ridden) (*6*).

Although earlier research has examined the crash and injury characteristics of PTW, mopeds are rarely the main focus of road safety studies (*2*, *7*, *8*). Additionally, moped crashes appear to be under-represented in popular media items about road crashes, which can lead to an underestimation of this issue with the wider audience and with road safety professionals (*9*).

Since the characteristics of mopeds and their riders, and the type of trips these vehicles are used for, are often very different from those of motorcycles, it can be expected that their crash characteristics differ substantially from those of motorcycles (*8*). Indeed, studies by Blackman & Haworth (*1*) and the European Association of Motorcycle Manufacturers (ACEM) (*10*) found substantial differences in crash patterns between mopeds and motorcycles.

[1]Vias Institute, Brussels, Belgium

Corresponding Author:
Address correspondence to Tim De Ceunynck: tim.deceunynck@vias.be

Despite their small share in many industrialized countries' modal split, mopeds seem a relevant and understudied topic in road safety research. The causal factors behind moped crashes are still largely unknown. Therefore, this study will focus on moped crashes; more specifically, the aim of this study is to get a better understanding of the causes and circumstances of road crashes with mopeds in urban areas. The main research questions are:

- What are the characteristics as well as human, vehicle-related, infrastructural and environmental crash factors of moped crashes in urban areas?
- What are the typical crash profiles commonly found in moped crashes in urban areas?

To this aim, an in-depth analysis of police reports of 167 severe injury moped crashes that took place in Belgium in 2013 was conducted. The next section provides a short background about earlier moped crash causation studies and about relevant moped legislation. The subsequent section describes the research methodology in more detail. The results of the data analyses are then presented, followed by a discussion of the findings. The final section reviews the main conclusions.

Background

Crash Causation

Earlier studies identified age and gender as relevant factors in moped crashes. In many countries, mopeds are popular with adolescents, for whom a combination of inexperience, risk-taking, and risk-exposure leads to a higher crash risk. Research suggests that the crash risk is highest for extreme age groups (i.e., the youngest and the oldest moped riders) (*5, 11*). Van Elslande & Marechal (*12*) found that young moped riders deliberately violated a fundamental traffic rule in one out of five crashes in the investigated sample. Male moped riders have a higher risk of being involved in a crash than female riders (*11*).

Moskal et al. (*11*) showed that riders not wearing a helmet have twice the risk of being injured when involved in a crash compared with helmeted riders. It was found that driving under the influence (DUI) of alcohol was the largest risk factor. Another finding from said study was that moped riders carrying a passenger had a higher crash risk than those without a passenger. Also, there is a higher crash risk when making a leisure trip than a home-to-work or home-to-school trip. Ignorance of the road during leisure trips can cause riders to be caught unawares by a deficiency in the infrastructure (such as poor surface condition) (*11*).

Research suggests that the operators of the other (non-PTW) vehicles are at fault in the majority of crashes between PTW and another motor vehicle (*6*). Often, the other driver involved fails to see and give way to the PTW, whether for reasons associated with the visual characteristics of the PTW or because of a lack of expectancy (*6, 10*). Drivers take little account of mopeds because there are relatively few mopeds on the road (*13*). Research shows that a motor vehicle driver had not seen the moped half of all crashes, which is in part caused by the fact that smaller vehicles are more difficult to perceive and their speed is more difficult to assess (*12*). Moped crashes are generally more severe at night (*1*).

Kosola et al. (*14*) found that 52% of moped injury crashes were crashes with other vehicles, while 33% were injured falling down. The remainder of the crashes included riding off the road, crashes with roadside objects, and with non-motorized other road users.

A study by Møller and Haustein (*7*) about crashes involving young moped riders found that 40% of the mopeds involved in the analyzed crashes were tuned up (i.e., the moped is tampered with so that it can ride faster than it is legally allowed to). In 29% of the crashes, the moped being tuned up was registered as a contributing factor in the crash.

Legislation

In Belgium, one has to be at least 16 years old to operate a moped, and at least 18 years to carry a passenger. All riders and passengers are required to wear a helmet, and each moped must be insured. Two categories of mopeds relevant to the purpose of this study are distinguished in Belgium: mopeds of class A and class B.

Class A mopeds are limited to a maximum speed of 25 km/h (16 mph). A driving license is not required to operate a moped class A. Mopeds class B have a maximum speed of 45 km/h (28 mph), and the rider is required to have a driving license (either a dedicated moped license or any other driving license). Class B mopeds are allowed to ride both on the road and on the bicycle path within built-up areas, and outside the built-up area they must ride on a bicycle path if there is one available. Class A mopeds, however, must always ride on the bicycle path when available (*15*).

A comparable classification into two subclasses of moped is made in some other European countries such as Germany, The Netherlands, Switzerland and the Scandinavian countries. However, most countries, including the United States, Canada, United Kingdom, and France, do not make such a subdivision.

Methodology

All severe injury crashes (i.e., at least one victim admitted to hospital for more than 24 hours) in Belgium in 2013 that took place in an urban area and involved a moped were selected from the official national injury crash database (*16*); a total population of 197 injury crashes. For all these crashes, the complete police report was requested from the responsible parquets. 167 of these police reports were obtained, which is a sample of 85% of the population.

The police report contains a detailed description of the general circumstances of the crash, the vehicles involved in the crash, drivers and passengers, and the infrastructure. The information from the police reports was registered in a predefined structured code book. The code book contains approximately 100 different variables and is based on the Initiative for the Global Harmonisation of Accident Data (IGLAD) framework (*17*), supplemented with several additional factors that were considered potentially interesting.

The coding of the police records was performed by coders who had received training to ensure that the data were coded accurately and consistently. Regular discussions and consistency checks between the coders took place during the entire coding process.

The data from the codebook were analyzed using descriptive statistics and significance tests where appropriate. The key results are presented in the following section.

After performing the statistical analyses of the data from the codebook, several crash profiles (i.e., homogeneous groups of crashes with several common core characteristics) were identified. First, individual crashes were compared with one another. Each crash was compared both to the already formed groups of crashes and to individual crashes that were not yet assigned to a group. When comparing crashes, each crash as a whole was considered, not just specific predefined details of the crash. Sometimes only one aspect of crashes was selected as the basis to form a crash profile, sometimes multiple aspects were considered simultaneously. A residual category remained that includes several crashes that have little or no similarities with other crashes. In this project, the crashes were first divided into crash profiles by two individual researchers working independently of each other. By comparing the results from both researchers and discussing the similarities and differences, a final list of crash profiles was compiled. The characteristics and causes of the crashes in each profile were then analyzed in more detail. This method is similar to the method developed by (*18*, *19*).

Analyses and Results

Characteristics of Moped Crashes

Vehicles and Drivers Involved. A total of 168 mopeds were involved in the 167 crashes analyzed. The dataset includes four fatal crashes. All fatally injured victims were moped riders. The proportions of mopeds of class A (49%) and class B (45%) in the sample were similar. For 5% of the mopeds involved the class was unknown, and one other type of moped (moped 3–4 wheels) was registered.

It is difficult to obtain reliable estimates of the exposure and/or full fleet of the different classes of moped, since registration of mopeds only became mandatory in Belgium in 2014, and the registration of vehicles sold before that date is still ongoing. Unpublished numbers of the registrations of new mopeds in 2015–2016 were obtained from direct communication with FEBIAC (Belgian Federation of the Car and Two-wheeler Industries). These numbers show that in 2015–2016, 10,329 class A mopeds were registered, and 14,359 mopeds of class B. In other words, the number of registered class B mopeds is nearly 40% higher than the number of class A mopeds. While it remains uncertain how closely these proportions represent the true fleet of mopeds as well as the total exposure to traffic in terms of mileage, this is an important indication that more kilometers are ridden on class B mopeds and thus that class A mopeds are over-represented in severe moped crashes in urban areas.

The ages of the moped riders involved are shown in Figure 1. The average age of all involved moped riders was 33 years. The age category 16–17 years was the largest category (21%), while 46% of all riders were within the range of 15–24 years old, which is a somewhat higher proportion than for all moped rider fatalities in the EU-24 countries (39%) (*20*). The age distribution was fairly similar for moped classes A and B, except for a substantially larger proportion of 40–49 year-old riders in the class A moped category, while the share of class B mopeds was larger for riders aged 16–17. The average age of the other (non-moped) drivers involved in the crashes was 40.

Of the moped riders involved 76% were male; 23% female. For 1% of the moped riders the gender was unknown.

The total share of single-vehicle moped crashes was 21%, which is similar to the share of single-vehicle crashes in the total population of all injury crashes in Belgium (*4*).

Infrastructure. Table 1 presents the infrastructural characteristics related to the moped crashes investigated. The

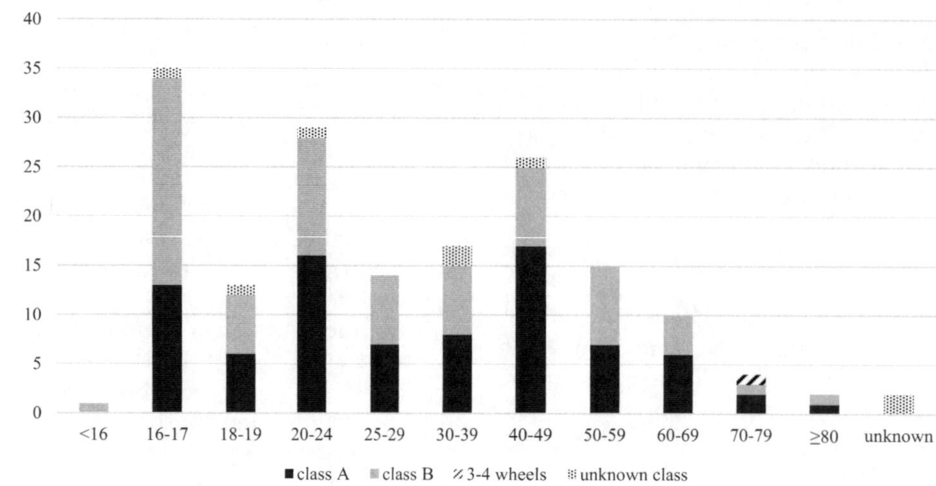

Figure 1. Age of moped riders involved.

analyses show a nearly equal distribution of the moped crashes investigated between road sections (51% of all crashes) and intersections (49%). Of all injury crashes that took place in Belgium in 2013 64.4% occurred on road sections, while 35.6% took place at intersections (*4*). Therefore, severe moped crashes within built-up areas occurred significantly more frequently at intersections compared with all types of injury crashes combined ($\chi^2(1) = 13.222$; $p < 0.001$).

Looking further into the type of intersection, it appears that the largest proportion (22%) of moped intersection crashes took place at T-shaped intersections, followed by four-legged intersections. The proportion of moped crashes at roundabouts is low in absolute terms (4%), but significantly higher than the share of all injury crashes that took place at roundabouts (1.2%) ($\chi^2(1) = 8.446$; $p = 0.004$) (*4*). The highest share of moped crashes took place at intersections managed by yield/stop signs (19%), followed by priority to the right intersections (12%).

Contributing Factors in Moped Crashes. A list of 247 possible factors contributing to moped crashes was composed. The list is an extension of a list that was composed for an earlier in-depth study about motorcycle crashes (*21*). The list consists of a three-level taxonomy. The highest level distinguishes between human factors, vehicle factors, infrastructural factors, and environmental factors. Each of these four categories is broken down into several subcategories, which in turn contain multiple individual contributing factors each.

Up to five contributing factors could be selected for each of the crash partners involved. A crash partner is defined as a vehicle or a pedestrian that is involved in a crash, thus a collision between a moped rider and a private car consists of two partners. In total, 342 factors were selected for the 299 crash partners that were involved in the 167 investigated crashes, which equals 1.14 factors per crash partner involved on average. The highest number of factors coded for an individual crash partner was four, the lowest was zero. An overview of the identified crash factors is presented in Table 2. Note

Table 1. Infrastructural Characteristics of Moped Crashes

Infrastructural characteristic	n	%
Intersection vs. road section		
Road section	85	51%
Intersection	82	49%
Type of intersection		
Not applicable (road section)	85	51%
T-intersection (three-legged, perpendicular angle between approaches)	37	22%
Y-intersection (three-legged, non-perpendicular angle between approaches)	7	4%
Four-legged intersection	24	14%
Roundabout	6	4%
Other	7	4%
Unknown	1	1%
Applicable right-of-way		
Not applicable (road section)	85	51%
Traffic lights (operational)	12	7%
Traffic lights (not operational)	1	1%
Yield/stop signs	32	19%
Priority to the right	20	12%
Other	9	5%
Unknown	8	5%

Note: n = number of moped crashes; % = percent of total moped crashes.

Table 2. Identified Factors Contributing to Moped Crashes

Crash factors (grouped)	Moped n	Moped %	Other road user n	Other road user %
Human				
Psychological factors	32	19.2%	38	22.8%
Risk assessment errors	34	20.4%	25	15.0%
Other violations	25	15.0%	16	9.6%
Loss of control over the vehicle	30	18.0%	0	0%
DUI of alcohol/drugs	26	15.6%	2	1.2%
Other risky behavior	10	6.0%	5	3.0%
Excessive or inappropriate speed	8	4.8%	2	1.2%
No helmet or worn incorrectly	4	2.4%	0	0%
Behavior of passenger	3	1.8%	1	0.6%
Lack of experience	2	1.2%	1	0.6%
Physical factors	1	0.6%	1	0.6%
Involvement in secondary tasks	0	0%	2	1.2%
Vehicle				
Tuned up vehicle	2	1.2%	0	0%
Sight obstruction due to vehicle design	1	0.6%	0	0%
Mechanical failure	1	0.6%	0	0%
Poorly secured load	1	0.6%	0	0%
Infrastructure				
Sight obstruction due to infrastructure	7	4.2%	9	5.4%
Grip	5	3.0%	0	0%
Roadworks	3	1.8%	1	0.6%
Issues regarding road lay-out	3	1.8%	1	0.6%
Poor condition of the pavement	2	1.2%	0	0%
Environment & circumstances				
Sight obstruction due to another vehicle	8	4.8%	15	9.0%
Congestion	2	1.2%	4	2.4%
Adverse lighting conditions and poor lighting	2	1.2%	1	0.6%
Low sun	0	0%	3	1.8%
Precipitation	2	1.2%	0	0%
Obstacle on the roadway	1	0.6%	0	0%

Note: n = number of moped crashes; % = percent of total moped crashes.

that, for reasons of brevity, the individual factors are not displayed, only the highest levels and their sublevels. Where relevant, individual factors are however mentioned in the interpretation in the following paragraphs.

Not surprisingly, human factors were by far the largest category of factors contributing to moped crashes. The largest subgroup of human factors constituted psychological factors; it was registered for the moped rider in 19.2% of the crashes, and in 22.8% of the crashes for the other road user involved. This category includes aspects such as inattention, distraction, hastiness and panic reactions. Risk assessment errors were a major subcategory as well, registered for the moped rider in 20.4% of the crashes and for the other involved road user in 15% of the crashes. Loss of control over the vehicle was registered for the moped rider in 18% of the crashes. DUI of alcohol or drugs was a factor registered for the moped rider involved in 15.6% of the crashes, and in 1.2% of the crashes for the other road user. It is noteworthy that excessive or inappropriate speed was only registered for a relatively low number of road users involved. The relatively low importance of excessive speed in these crashes is in line with (*10*), but not in line with (*7*) who found that riding speed was the most frequently assigned factor on the part of the moped rider. Wearing no helmet or wearing it incorrectly was noted as a contributing factor for four moped riders (2.4% of the crashes).

Vehicle-related factors only played a minor role in the crashes investigated. The factor "tuned up" is however worth some more explanation. In total, 11 mopeds involved (six class A mopeds and five class B mopeds) were found to be tuned up, and in two crashes this was registered as a factor that actually contributed to the crash. However, only 19 out of 168 mopeds were tested. Some mopeds were too severely damaged to be tested after the crash, and not all police forces had the necessary equipment available to test the moped's performance. Taking into account the findings by Møller and Haustein (7), it is likely that the importance of this factor is underestimated, although it must be mentioned that said study focused on the youngest (16–17 year old) moped riders, who might be more inclined to tune up their moped than other age groups. In our database, six out of 11 mopeds that were tuned up were ridden by a rider aged 16–17.

The most common infrastructural factor is the presence of a sight obstruction due to an infrastructural element, which is registered for the moped rider in 4.2% of the crashes and for the other road user involved in 5.4% of the crashes. In 3% of the crashes, the moped rider experienced issues with grip. The presence of roadworks, issues related to the road lay-out, and the condition of the pavement were registered in a limited number of crashes.

By far the most common environmental factor is sight obstruction due to another vehicle, which was registered for the moped rider in 4.8% of the crashes and for the other road user in 9% of the crashes. The presence of congestion, adverse lighting conditions and poor lighting, low position of the sun, the presence of precipitation, and an obstacle on the roadway were registered in a limited number of crashes as well.

Most Common Crash Profiles and Their Characteristics. Table 3 provides an overview of the crash profiles identified. The two most common crash types were crashes in which a motor vehicle driver crashed into a moped rider while turning left or right, and crashes caused by a moped rider's risky behavior.

In the following, each crash profile is discussed, except for the 'other' category for which no useful conclusions could be drawn as it was too heterogenous. A description of the crash situation is provided and the most common crash factors are discussed. Where relevant, results of a comparison between the crashes in that crash profile with the crashes of all other profiles regarding the types of moped involved and the age of the moped riders are reported.

Table 3. Overview of Crash Profiles

Profile	n	%
Crashes between a vehicle turning off and a moped going straight through	30	18 %
Crashes caused by the moped rider's risky behavior	28	17 %
Crashes between a moped and another vulnerable road user (VRU)	22	13 %
Crashes resulting from entering/exiting destinations alongside the road	20	12 %
Crashes at intersections (other than vehicles turning off)	18	11 %
Single-vehicle crashes with loss of control	15	9 %
Crashes caused by an error during overtaking	13	8 %
Rear-end crashes	6	3 %
Other	15	9 %

Note: n = number of moped crashes; % = percent of total moped crashes.

Profile 1: Crashes between a Vehicle Turning Off and a Moped Going Straight Through (30 Crashes)

There are two subtypes in this crash profile; one where a motor vehicle turned right and collided with a moped riding in the same direction on a bicycle path or on the road (19 crashes), and one where the motor vehicle turned left and crashed with a moped riding in the opposite direction on a bicycle path (11 crashes). Ten of the right-turn crashes took place on a one-way bicycle path, with all but one of the mopeds riding in the normal direction. The other nine right-turn crashes took place on the roadway. All of the crashes involving a left-turning vehicle took place on a bicycle path; six on a one-way bicycle path and five on a two-way bicycle path. Nine moped riders were riding in the normal direction, two were riding in a counterflow direction (one of them was riding in a counterflow direction on a one-way bicycle path).

In this crash profile, risk assessment errors and inattention were the most common behavioral factors, which were registered 18 times (seven times for the moped rider, 11 times for the other vehicle) and 16 times (nine times for the moped rider, seven times for the other vehicle), respectively. Sight obstructions are by far the most common non-human factor, registered for three moped riders and for eight of the other road users.

Profile 2: Crashes Caused by the Moped Rider's Risky Behavior (28 Crashes)

This profile includes various types of crash in which it is clear that risky behavior of the moped rider was a major factor in the occurrence of the crash. The most common scenario in this crash profile is a moped rider DUI who collided with an object, or a parked or driving vehicle, or who simply fell down (18 crashes in total). The risky behavior can also be excessive speed, stunting with the vehicle, etc.

DUI and loss of control were the most common crash factors in this profile, followed by other violations, risk assessment errors, and not wearing a helmet or wearing it incorrectly. Vehicle, infrastructure-related and environmental factors were rarely coded for crashes in this profile. In two cases, the fact that the moped had been tuned up was a contributing factor.

Somewhat surprisingly, the moped riders involved in these crashes were not generally younger than in other crash types ($t(163) = 1.108$; $p = 0.269$). The (non-significant) best estimate even suggests that moped riders in crashes caused by risky behavior were somewhat older than moped riders in other crash profiles. Of the crashes in this profile 36% took place during the night (between 10.00 p.m. and 5.59 a.m.), which is a significantly higher proportion than for the other crash profiles ($\chi^2(1) = 6.546$; $p = 0.011$).

Profile 3: Crashes between a Moped and Another Vulnerable Road User (VRU) (22 Crashes)

We can distinguish two common subtypes within this profile: a moped riding on a bicycle track or on the road that crashed with another VRU that wanted to cross the road (usually a pedestrian) (eight crashes); and a moped riding on the bicycle track that crashed with a cyclist riding on the same track in the opposite direction (six crashes).

The most common crash factors were behavioral factors. Risk assessment errors, inattention and committing a violation were most commonly coded. In the majority of cases, these factors were coded for the other VRU.

The proportion of class A mopeds was significantly lower (and consequently the percentage of class B mopeds was higher) than for other crash profiles ($\chi^2(1) = 4.251$; $p = 0.039$). In one way, this could be seen as unexpected because the required position on the road of class A mopeds (i.e., always on the bicycle path, when present) could lead to a higher number of interactions with other VRU (especially cyclists). On the other hand, the larger speed differences between class B mopeds and other VRU might contribute to a higher number of (and possibly more severe) interactions with other VRU as well.

Profile 4: Crashes Resulting from Entering/Exiting Destinations alongside the Road (20 Crashes)

The vast majority of these crashes (14 crashes) took place when a vehicle entered or exited (the parking lot of) a destination alongside the road, such as a store, petrol station or driveway. In only two crashes was it the moped rider who exited the adjacent destination; in all other crashes it was the other motor vehicle driver who entered or exited the adjacent destination. Four crashes took place when a vehicle parked in parallel pulled out. Two crashes were somewhat different and occurred when an occupant of a vehicle parked in parallel carelessly opened the door, hitting a passing moped rider.

Inattention was coded as a contributing factor most frequently (15 times). In the vast majority of cases (12 times), this factor was coded for the other road user involved. Sight obstructions by the infrastructure and by the environment were the most common non-human crash factor (registered six times).

Profile 5: Crashes at Intersections (Other Than Vehicles Turning Off) (18 Crashes)

Similar to profile 1, this profile constitutes crashes that took place at intersections, but the difference is that the crashes in profile 5 were not caused by turning manoeuvers. They either constitute situations in which a moped was riding on the roadway and both road users arrived approximately simultaneously at the intersection (10 crashes), or in which the moped was riding on a bicycle path on the priority road and was hit by a car entering from a subordinate road (eight crashes). These crashes usually took place at four-leg intersections (13 crashes). It is noteworthy that the moped rider had the right-of-way in the majority of these crashes; the moped rider had the right-of-way in 12 crashes, the other road user involved in three crashes, and in three crashes the police report did not enable us to identify who had the right-of-way. Inattention, risk assessment errors, and sight obstructions were common crash factors.

Profile 6: Single-Vehicle Crashes with Loss of Control (15 Crashes)

This profile includes a variety of crashes in which a moped rider lost control over the vehicle without the involvement of another road user and crashed, for example, because the rider misjudged a curve, braked too hard, or was affected by the state of the road, shifting luggage, etc. The rider fell down or hit an obstacle such as a pole, another vehicle (which is considered a secondary crash here), or a wall. Of these crashes 40% took place at night, which is a significantly higher proportion than for other crash profiles ($\chi^2(1) = 5.010$; $p = 0.025$). Only two of the crashes in this profile took place on a bicycle path; the vast majority of the crashes took place on the roadway. Only one of these crashes occurred at an intersection, all other crashes took place at a road section. The analysis of registered crash factors could not highlight any predominant contributing factors. The most common factor was issues with grip (registered three times).

Profile 7: Crashes Caused by an Error during Overtaking (13 Crashes)

In this category, it was usually the moped rider who performed the overtaking manoeuver. Only in two crashes did an overtaking motor vehicle hit a moped. For the remaining crashes, in which the overtaking manoeuver was performed by the moped rider, four subcategories are distinguished: 1) a motor vehicle initiated a left turn while being overtaken by a moped (four crashes); 2) for various reasons, the moped rider lost control during overtaking (four crashes, twice overtaking on the left-hand side, twice on the right-hand side), for example by hitting a curb; 3) the moped rider overtook another road user on the bicycle path and crashed into a cyclist coming from the opposite direction (two crashes; since both crashes took place on a two-way bicycle path, the opposing cyclists were riding in the correct place); 4) one crash in which a moped overtook a passenger car and crashed into a car driving in the opposite direction.

The most common human crash factor in this profile was a violation (registered three times for the moped rider, seven times for the other road user involved). Most frequently, the violation was not indicating a manoeuver (e.g., a left turn). The average age of the moped riders in this crash profile was only 21.5 years, which was significantly younger than the average age of 33.9 that was found in all other crash profiles combined ($t(163) = 2.56$; $p = 0.012$).

Profile 8: Rear-End Crashes (6 Crashes)

In four of the rear-end crashes, it was the moped rider who crashed into the rear of a stationary or braking vehicle, while in two crashes a stationary moped waiting at an intersection was rear-ended by a motor vehicle. Five crashes involved a class A moped and one involved a moped of unknown class. No class B mopeds were involved. The significance test suggests that the share of mopeds class A was marginally significantly higher than for the other crash profiles ($\chi^2(1) = 0.624$; $p = 0.060$), but due to the low number of crashes in this category, this result should be interpreted with caution.

Discussion

Strengths, Weaknesses, and Future Research

An important strength of the study is that the dataset contains 85% of all eligible moped crashes in Belgium in 2013, which can be considered a very representative sample. From the perspective of an in-depth study, a sample of 167 crashes is a relatively large dataset. A limitation is however that, in statistical terms, this remains a relatively low number of records, which poses some challenges for drawing firm statistical conclusions, especially regarding contributing factors and crash characteristics that are relatively uncommon.

Another strength of the study is that the use of in-depth data from police reports provides much richer data than traditional crash studies that make use of police-recorded crash data. An important limitation is however that, from a research point of view, the police reports are not always complete. Some data are systematically missing, such as seat belt use, the age of the vehicle and the result of the alcohol blood test. It is also impossible to identify all contributing factors based on an analysis of police reports. Quite a number of these factors remain hidden because there is not enough information in the file. In-depth research with on-site crash investigations provides additional possibilities to identify causal factors of the crashes and is therefore a useful direction for future research. In addition, it is known that the content of police reports can be subject to substantial biases in content. A police report is always a summary of an interrogation and investigation in which information is left out, summarized, reformulated and/or added, which can introduce biases (*22*). Other potential sources of bias are the emergence of tunnel vision during the registration or as a result of prior knowledge (*23*). Other possible biases to which these data might be susceptible are related to the main goal of the data collection (assigning culpability rather than identifying all contributing factors that might be useful from a policy perspective), and perception biases and attitudes toward mopeds held by the assessor.

Dividing the crashes into several crash profiles has allowed us to discern some dominant patterns in the data, and to analyze the contributing factors in a more targeted way. The process has been made as objective as possible by using a proven methodology, by making maximum use of the rich data and earlier analyses, by having two researchers making the distribution independently in the first stage, and by extensive discussion about the final profiles and subdivision. Nevertheless, the identification of common patterns remains a qualitative process that is to some extent open to discussion. A limited number of crashes could be put in more than one category, and in such cases it can be debated which feature(s) should prevail.

Since little research has been conducted into moped safety, further research is needed. Transferability of the results cannot be ensured, especially not to countries that have a transport context that is strongly different from Belgium. When using the results of this study, it is important to keep in mind that the study took place in a highly motorized, densely populated country with generally high volumes of motorized traffic, but low volumes of mopeds.

Policy Implications

The findings of this study have several implications for policy makers. Of the crashes 91% could be divided into a limited number (eight, more specifically) of rather similar categories for which targeted measures can be designed or specific points of attention taken into account to reduce their frequency.

Implications for Road Design

The most common crash profile (profile 1) were crashes at intersections involving a left or right turning vehicle. Sight obstructions (caused by infrastructural elements or by other vehicles) were a common factor in these crashes. Also in some of the other profiles (profile 4, entering and exiting destinations alongside the road; and profile 5, non-turning intersection crashes) this was to some extent observed. Avoiding obstacles that obstruct lines of sight is therefore an important point of attention in infrastructure design to reduce the number of severe moped crashes in urban areas.

Single-vehicle crashes with loss of control were a substantial category of moped crashes as well. It is not straightforward to pinpoint the most promising interventions to reduce these crashes, since they can be caused by errors on the part of the driver, infrastructural deficiencies, unfortunate conditions or situations, or a combination. Moreover, it is well-known that single-vehicle VRU crashes are strongly under-reported in police-reported crash statistics and their causes are not always reliably registered (*24*). Therefore, the importance of this category of crashes might be underestimated. A study (*25*) on single-bicycle crashes suggests that infrastructural factors contribute to approximately half of those crashes. Even though the importance of infrastructural elements could not be confirmed as strongly from our analyses of the police reports, it is possible that details in design or minor flaws in the state of the infrastructure that were considered "normal" by the assessor still contributed to the moped rider's loss of control. Attention to the detailed road design and the condition of the infrastructure to avoid obstacles or deficiencies that could cause a fall therefore seems to have a high potential to reduce the number of crashes of this type. Road safety audits and inspections (specifically from the perspective of moped riders) are a promising tool to avoid at least some of these crashes. The high share of night-time occurrence in single-vehicle crashes is an important point of attention in these audits and inspections.

Implications for Education and Enforcement

Since psychological factors and risk assessment errors were quite common in most of the moped crash profiles, sensitization and/or education might help to reduce moped crashes. Since these factors were also registered frequently for the other (non-moped) road user involved, these measures should be targeted not only t moped riders, but also at other drivers.

Young moped riders were over-represented in overtaking crashes (profile 7), which somewhat surprisingly constituted a rather large group of crashes. Even though overtaking as such is not necessarily a risky behavior, the line between a safe and an unsafe overtaking manoeuver can be thin. Risk assessment errors and overconfidence can strongly contribute to these crashes. Driver education could contribute to raising awareness of the risks of overtaking, as well as to training (young) moped riders to overtake more safely.

In line with the study by Van Elslande & Marechal (*12*), crashes caused by risky behavior of the moped riders (profile 2) were an important crash category as well. Risky behavior should therefore be considered a major issue to moped safety. Targeted enforcement and awareness-raising could help to mitigate crashes in this category. Somewhat surprisingly, younger moped riders were not over-represented in crashes caused by risky behavior.

Implications for Legislation

Quite remarkably, the third largest category were crashes between a moped and another VRU (profile 3). A higher share of crashes involving mopeds of class B was found in this category, possibly because their higher speed could induce more severe crashes. This finding could open a debate on whether mopeds of class B should be allowed to make use of cycle paths that are mostly used by slower VRU. In particular, a number of the crashes with cyclists would probably not have taken place if the moped riders had been required to use the roadway instead of the bicycle path (although it should be taken into account that crashes between moped riders and other motor vehicles might increase in this case). Given the increasing variety in characteristics and speed of vehicles that make use of facilities for VRU (e.g., pedestrians, bicycles and mopeds, but also electric bikes and speed pedelecs, "new urban mobility modes" such as monowheels and electric steps, cargo bikes,...), the position and rules of conduct for each of these categories will become an important point of debate in any case.

Conclusion

The proportion of mopeds in both class A and B in severe moped crashes in urban areas is approximately fifty-fifty, which seems to imply an over-representation of mopeds of class A in these crashes. The moped riders involved have a mean age of 33 and three-quarters are men.

Human factors are by far the largest category of contributing factors in moped crashes, factors related to infrastructure and environment play a moderate role, and vehicle-related factors play only a minor role. The most important human factors are psychological factors and risk assessment errors, the main non-human contributing factors are sight obstructions due to infrastructure or other vehicles.

Nine crash profiles, grouping similar types of crashes together, were distinguished:

- Crashes between a vehicle turning off and a moped going straight through (18%)
- Crashes caused by the moped rider's risky behavior (17%)
- Crashes between a moped and another VRU (13%)
- Crashes resulting from entering/exiting destinations alongside the road (12%)
- Crashes at intersections (other than vehicles turning off) (11%)
- Single-vehicle crashes with loss of control (9%)
- Crashes caused by an error during overtaking (8%)
- Rear-end crashes (3%)
- Other (9%)

Author Contributions

The authors confirm contribution to the paper as follows: study conception and design: De Ceunynck, Slootmans & Daniels; data collection: Slootmans; analysis and interpretation of results: De Ceunynck, Slootmans & Daniels; draft manuscript preparation: De Ceunynck. All authors reviewed the results and approved the final version of the manuscript.

References

1. Blackman, R. A., and N. L. Haworth. Comparison of Moped, Scooter and Motorcycle Crash Risk and Crash Severity. *Accident Analysis and Prevention*, Vol. 57, 2013, pp. 1–9. https://doi.org/10.1016/j.aap.2013.03.026.
2. SafetyNet. *Powered Two Wheelers: Web Text.* 2009. https://ec.europa.eu/transport/road_safety/sites/roadsafety/files/specialist/knowledge/pdf/powered_two_wheelers.pdf. Accessed July 31, 2017.
3. European Commission. *Traffic Safety Basic Facts 2016 on Motorcycles & Mopeds.* European Commission, Directorate General for Transport, Brussels, Belgium, 2016.
4. Schoeters, A. *Statistisch Rapport 2014 Verkeersongevallen in Dutch.* Publication R2015-S-04-NL. Belgian Road Safety Institute: Knowledge Center Road Safety, Brussels, Belgium, 2014.
5. Martensen, H. *@Risk - Analyse van het risico op ernstige en dodelijke verwondingen in het verkeer in functie van leeftijd en verplaatsingswijze [in Dutch].* Publication D/2014/779/37. Belgian Road Safety Institute, Knowledge Center Road Safety, Brussels, Belgium, 2014.
6. Haworth, N. Powered Two Wheelers in a Changing World - Challenges and Opportunities. *Accident Analysis and Prevention*, Vol. 44, No. 1, 2012, pp. 12–18. https://doi.org/10.1016/j.aap.2010.10.031.
7. Møller, M., and S. Haustein. Factors Contributing to Young Moped Rider Accidents in Denmark. *Accident Analysis & Prevention*, Vol. 87, 2016, pp. 1–7. https://doi.org/10.1016/j.aap.2015.11.008.
8. White, D., J. Lang, G. Russell, K. Tetsworth, K. Harvey, and N. Bellamy. A Comparison of Injuries to Moped/Scooter and Motorcycle Riders in Queensland, Australia. *Injury*, Vol. 44, 2013, pp. 855–862.
9. De Ceunynck, T., J. De Smedt, S. Daniels, R. Wouters, and M. Baets. "Crashing the Gates": Selection Criteria for Television News Reporting of Traffic Crashes. *Accident Analysis and Prevention*, Vol. 80, 2015, pp. 142–152. https://doi.org/10.1016/j.aap.2015.04.010.
10. ACEM. *MAIDS: In-depth Investigations of Accidents Involving Powered Two Wheelers, Final Report 2.0.* European Association of Motorcycle Manufacturers, Brussels, Belgium, 2009.
11. Moskal, A., J.-L. Martin, and B. Laumon. Risk Factors for Injury Accidents Among Moped and Motorcycle Riders. *Accident Analysis and Prevention*, Vol. 49, 2012, pp. 5–11.
12. Van Elslande, P., and M. Marechal. *Accidentologie des cyclomoteurs, Rapport final [in French].* IFSTTAR, Unité de recherche Mécanismes d'accidents, Salon-de-Provence, France, 2008.
13. Noordzij, P., E. Forke, R. Brendicke, and B. Chinn. *Integration of Needs of Moped and Motorcycle Riders into Safety Measures. Review and Statistical Analysis in the Framework of the European Research Project PROMISING, Work Package 3.* SWOV Institute for Road Safety Research, Leidschendam, the Netherlands, 2001.
14. Kosola, S., P. Salminen, and T. Laine. Heading for a Fall - Moped and Scooter Accidents from 2002 to 2007. *Scandinavian Journal of Surgery*, Vol. 98, No. 3, 2009, pp. 175–179.
15. Wegcode, be. *Koninklijk besluit van 1 december 1975 houdende algemeen reglement op de politie van het wegverkeer en van het gebruik van de openbare weg [in Dutch].* https://wegcode.be/wetteksten/secties/kb/wegcode. Accessed March 31, 2017.
16. Statistics Belgium. Accidents et victims de la circulation [in French]. http://economie.fgov.be/fr/statistiques/chiffres/circulation_et_transport/circulation/accvict/. Accessed November 14, 2017.
17. IGLAD. *Initiative for the Global Harmonisation of Accident Data.* http://www.iglad.net/. Accessed July 31, 2017.
18. Brenac, T., and D. Fleury. Le concept de scénario type d'accident de la circulation et ses applications [in French]. *Recherche Transport Sécurité*, Vol. 63, 1999, pp. 63–76.
19. Brenac, T., and B. Megherbi. Diagnostic de sécurité routière sur une ville: intérêt de l'analyse fine de procédures d'accidents tirées aléatoirement [in French]. *Recherche Transports Sécurité*, Vol. 52, 1996, pp. 59–71.
20. Broughton, J., C. Brandstaetter, G. Yannis, P. Evgenikos, P. Papantoniou, N. Candappa, M. Christoph, K. van Duijvenvoorde, M. Vis, J.-F. Pace, M. Tormo, J. Sanmartín, M. Haddak, L. Pascal, E. Amoros, P. Thomas, A. Kirk,

and L. Brown. *Assembly of Basic Fact Sheets and Annual Statistical Report - 2012*. Deliverable D3.9 of the EC FP7 project DaCoTA, 2013.
21. Martensen, H., and M. Roynard. *MOTAC - Motorcycle accident causation. Diepteanalyse van zware en dodelijke ongevallen waarin motorfietsers betrokken waren [in Dutch]*. Publication D/2013/0779/14. Belgian Road Safety Institute, Knowledge Center Road Safety, Brussels, Belgium, 2013.
22. Kranendonk, R., H. Elffers, M. Komter, M. de Boer, M. Malsch, and J. de Keijser. PV's en opnamen van verdachtenverhoren: een wereld van verschil [in Dutch]. *Het Tijdschrift voor de Politie*, Vol. 77, No. 3/15, 2015, pp. 6–12.
23. van den Eeden, C. A. J., D. J. de Poot, and P. J. van Koppen. Forensic Expectations: Investigating a Crime Scene with Prior Information. *Science and Justice*, Vol. 56, No. 6, 2016, pp. 475–481.
24. Schepers, P., N. Agerholm, E. Amoros, R. Benington, T. Bjørnskau, S. Dhondt, B. De Geus, C. Hagemeister, B. P. Loo, and A. Niska. An International Review of the Frequency of Single-Bicycle Crashes (SBCs) and their Relation to Bicycle Modal Share. *Injury Prevention*, Vol. 21, 2014, pp. e138–e143. https://doi.org/10.1136/injuryprev-2013-040964.
25. Schepers, P., and K. K. Wolt. Single-Bicycle Crash Types and Characteristics. *Cycling Research International*, Vol. 2, No. 1, 2012, pp. 119–135.

The Standing Committee on Motorcycles and Mopeds (ANF30) peer-reviewed this paper (18-03999).

Article

TRR
JOURNAL OF THE TRANSPORTATION RESEARCH BOARD

Transportation Research Record
2018, Vol. 2672(34) 96–105
© National Academy of Sciences:
Transportation Research Board 2018
Article reuse guidelines:
sagepub.com/journals-permissions
DOI: 10.1177/0361198118797825
journals.sagepub.com/home/trr
SAGE

Risky Ride or Carefree Drive? An Analysis of Virginia's Moped Safety and Registration Law on Operator Collision and Injury Outcomes

Joseph Seymour[1]

Abstract

The U.S. State of Virginia enacted a two-phased moped operator law, SB 1038, in 2013 that was intended to reduce collisions and injuries by requiring basic safety equipment, possession of identification (ID), and obtaining vehicle registration. This study evaluates the law's effect on moped collisions, injuries, and fatalities through a two-part evaluation framework of the pre- and post-policy implementation periods. These components include a series tool, the interrupted time series analysis (ITSA), and a logistic regression of individual-level crash characteristics. The research design controls for gasoline prices, socioeconomic status, roadway conditions, population, seasonality, alcohol consumption, age, and gender, among other factors. Findings from ITSA demonstrate that the implementation of a helmet and ID requirements for moped operator reduced collision rates in the months following that requirement, however no such effect was found for reductions in serious injury and fatality rates, or frequencies following that stage and the vehicle registration requirement (referred to as Phase II). The logistic regression analysis of individual moped collision outcomes similarly found no support for Phase II's role in reducing the risk of serious injury and fatal outcomes. As a public safety policy, SB 1038 was partially effective through its reduction of moped collisions, though the mechanism for achieving this is unclear. Potential explanations include the policy's role in reducing the pool of risk-taking individuals through disincentivizing moped use, and mainstreaming safer moped operation. The study's results indicate that states without moped operator requirements could experience reduced moped collisions from adoption of similar legislation.

Not all road users experience the same level of safety. The multimodal diversity of the nation's urban and suburban areas is beneficial for economic resilience and expressing personal preferences, but transportation policy and infrastructure do not necessarily work toward the safety of all. Vulnerable road users (VRU) include pedestrians, cyclists, and exposed riders such as those on motorcycles, scooters, and mopeds. These users often encounter roadway infrastructure that is not designed primarily for them in relation to travel speed and facilities (i.e., separate on-road), which may create a greater risk of collision, injury, or death. Given changing attitudes and attention toward personal mobility and transportation accessibility, it is essential that state transportation investments and policies support safety for all users.

Motorcycles, a broad group of vehicles that encompasses off-road motorcycles, scooters, mini-bikes, and mopeds, experience disproportionate fatality and injury rates compared with other road users. In 2013, motorcyclists accounted for 14% of all traffic fatalities, despite representing only 3% of registered vehicles and 0.7% of all vehicle miles traveled (VMT) (*1*). In addition, motorcyclists were more 26 times more likely than passenger car occupants to die in motor vehicle crashes. This substantially increased relative risk of death compared with automobile occupants may be due to a combination of factors, including physical exposure of the rider themselves and inferior safety equipment (e.g., helmet compared with lap seat belt and airbags). Although fatalities for motorcyclists overall declined by 6% in 2013, fatalities from moped use (a subgroup of motorcycles with engines less than 50 cubic centimeter (CC) piston displacement capacity) increased by roughly 2% (*2*). The increasing nationwide trend of moped fatalities

[1]Vanasse Hangen Brustlin, Inc, Durham, NC

Corresponding Author:
Address correspondence to Joseph Seymour: josephfseymour@gmail.com

is alarming: from 103 in 2010 to 147 in 2015. This loss of life has coincided with the passage of legislation in several states aimed to improve moped-user safety through vehicle registration and operator safety requirements. In the summer of 2013, Virginia enacted Senate Bill 1038, a two-phased measure that required moped operators to wear helmets and carry government-issued identification (ID) effective July 1, 2013, followed by registering their moped with the Virginia Department of Motor Vehicles (DMV) by July 1, 2014.

This study proposes to evaluate the effect of Virginia's (VA) moped operator law on collision rates and the risk of serious injury and death to moped operators. The research design is a two-pronged research design consisting of the quasi-experimental interrupted time series analysis (ITSA) and logistic regression methodologies. The first component is ITSA, which compares levels of collisions, serious injuries, and fatalities in VA along during the pre- and post-policy periods of January 2010 to July 2013 and July 2013 to July 2016, respectively.

The second tranche is a logistic regression analysis of the relative risk of serious injury and fatality during VA's pre- and post-policy environment. Together, these two analyses will approach the question of whether—and to what extent—VA's regulatory actions reduced moped-involved collision rates and the odds of serious injury or the death of the moped operator. The outcome of the analysis will shed light onto the effectiveness of traffic and public health policy where application may be extended to those states without such moped user requirements.

As a class of vehicle, mopeds remain largely overlooked in the realm of traffic safety policy. Falling below the class of motorcycle in both travel speeds and size, but above that of the bicycle, the moped has operated in the gray area of transportation. A moped is defined in VA state statute as a vehicle with up to a 50CC or equivalent engine size and incapable of exceeding 35 mph (among other criteria). Confident estimates for state and national-level sales are non-existent, as are population estimates for existing fleet counts in states that do not require moped registration. Furthermore, the majority of academic literature on moped-user safety and moped-specific crash predictors is limited to Australian and European contexts and populations where such vehicles are more common (*3*). Comparable studies on motorcycle collision factors and nascent moped research are explored below to establish variables for consideration.

The limited literature on moped collision risk factors highlights the influence of alcohol and lack of licensing. In a cohort study of North Carolina emergency room admissions records, admitted patients from moped collisions had a relative risk of 4.8 of having a DWI (driving while intoxicated) compared with that of admitted motorcycle operators (*4*). In a review of Florida crash incidents, moped operators who were found to have consumed alcohol had a relative risk of 2.09 of experiencing a severe crash versus a non-severe crash (*5*). An elevated blood alcohol level has long been known to increase the risk of vehicular collision, however, with regards to mopeds, the presence of alcohol may indicate risk-taking behavior arising from the lack of licensing and alcohol prohibitions required for motorcycle and automobile operators.

Like alcohol consumption, license status is associated with increased collision risk. The lack of an operator's license has been associated with higher motorcycle and moped collision risk when controlling for age, vehicle ownership status, and crash type (*6*, *7*). Though a license itself may not insulate a moped operator from a collision, the process for obtaining one may preclude an individual prone to risky behavior (i.e., DWI) from legally operating a moped. Thus, risk reductions from licensing laws may have less to do with improving the operator's performance than reducing the pool of risk-prone operators.

Additional noteworthy moped risk factors include roadway design and speed. Higher speed roadways and those with four lanes or more have both been correlated with increased moped crash risk (*5*). Given that the top speed of mopeds is compatible with local roadways (<35 mph), there is elevated risk on roadways where both speeds and volumes exceed this vehicle class's native conditions. Rural roadways—with their higher speed limits and greater speed differential with mopeds—may increase crash risk and related injury.

Although presence of a license has been correlated to lower risk of moped and motorcycle collisions, the licensing programs themselves have shown varied levels of effectiveness for crash reduction. Legislative policies such as mandatory motorcycle licensing and automobile Graduated Licensing (GDL) programs have been shown to reduce collisions to a greater extent than state-run educational and voluntary operator efforts, even after controlling for economic activity and fuel price volatility (*8*, *9*). Yet, governmental policies that require an operator license and requisite training have shown limited safety improvement, with benefits accruing dependent on the age of the trainee and miles of roadway experience (*10*, *11*). A meta-analysis of motorcycle licensing programs found that such training may function as an impediment to legally operating a motorcycle, thus indirectly reducing crash risk through a reduction in exposure (*12*). Improvements to study design have been recommended to determine whether the training itself or its function as a gateway restricting unsafe operators is related to collision reduction.

The state of VA sought to accomplish three objectives through passage of its moped law, SB 1038. First, the state intended to recognize and mainstream moped users and vehicles through their inclusion in titling and registration requirements. This formality would bring the state in line with approximately 36 states—at that time—that required the registration of mopeds (*13*). Second, the legislation would address broad safety concerns of the public, law enforcement, legislature, and traffic safety advocates. Third, the state desired to support the work of law enforcement in returning stolen mopeds through the requirement of titling and registration. Operator insurance requirements, specific operator licenses, and high-speed roadway limitations were not pursued due to concerns over rider economic and mobility equity as an alternative to the private automobile.

Present in all three of the state's main justifications for the moped user law was the desire to know the population of moped users and its risk profile. Whereas automobile licenses and VMT are tracked and compared for automobile safety reporting, no tools or requirements were applied to moped riders in VA (though moped drivers would have to carry ID). Requiring registration—it was reasoned—would support future analysis of moped-user safety without specialized ID: the state could track the level of vehicles on the road to standardize future crash rates for collisions and serious injuries. However, assessing moped-user safety along a traditional automobile-centric framework would necessarily exclude consideration of the state's safety efforts prior to registration (given that the moped population would only be known confidently years following the law's implementation). The incorporation of other standardization and analytical methods could bridge the perceived gap in evaluating the law's effectiveness from the pre- and post-implementation periods.

Methodology

This study incorporates a two-part framework for evaluating the effectiveness of VA's moped operator law in reducing moped crash rates and the relative risk of serious injury and death. These parts include an ITSA of moped collision and injury frequencies and rates in VA pre- and post-policy, and a logistic regression analysis of VA moped crash outcomes across numerous policy and operator predictors. The inclusion of crash predictors is informed by those identified in moped and motorcycle literature as well as those inferred by VA in its justification of the moped user law itself. Taken altogether, these methods will provide additional robustness on the nature and direction of the effectiveness of VA's moped-user law in the absence of rider population and registration data.

Data Sources

The data sources for the two-part analysis combine state collision databases and federal measures of energy costs, population, and socioeconomic standing. The primary source of VA's population of moped collisions is the state's Department of Transportation's (VDOT) publicly accessible crash database. Recorded collisions are those that have occurred on public roadways, were reported to police, resulted in death or injury, property damage estimated above US$1,500, or a combination of these . Many aspects of the crash incidents are recorded, including location, setting (urban vs. rural), date and time of day, condition of the driver, vehicle type, initial conclusion on role of alcohol in crash contribution, driver demographic information, and much more.

Datasets on the population, wages, and fuel prices complement the collision and moped-user profiles from the VDOT crash databases. For population estimates, a combination was used of the U.S. census's decennial state and county population estimates for 2010 and VA's intercensal estimates for the years 2011 to 2016. The Energy Information Administration's mid-Atlantic general unleaded average monthly retail gasoline prices were included for the study period 2010 to 2016 and adjusted for inflation. Finally, average weekly wage data was gathered from the U.S. Bureau of Labor Statistics' Quarterly Census of Employment and Wages for VA counties and cities from Q1 2010 through Q2 of 2016. Although median measures of income would reduce the influence of outliers, aggregating per capita wages at the county level would both reduce this effect and allow consideration of economic data beyond the window of annual American Community Survey income measures.

Reliance on police crash reports alone for analysis of collision and policy outcomes is not without its drawbacks regarding several issues. Given that police reports are only generated for events on public roadways that were also reported to police, there is a risk of underreporting. Additionally, inspection of the state crash datasets revealed instances of person-level incompleteness and coding errors. Where crash details were omitted or coding was incorrect (i.e., seat belt noted for moped operator when the FR300 codebook requires a "Not Applicable" designation), null values were recorded. Analysis of crash data could be bolstered through the pairing of hospital room admission data to provide collision outcomes on private and unreported public crashes as well as more detailed information on alcohol consumption and injury severity. However, due to data availability, the use of hospital records is beyond the scope of this study.

Interrupted Time Series Analysis

This study's ITSA design is executed through the Stata® command "itsa" (*14*). When utilized for a single group analysis, such as the first prong of this design, ITSA projects the pre-intervention trend into the treatment period and serves as the counterfactual. This allows for the targeted evaluation of the moped policy on collision and injury outcomes. ITSA requires at a minimum four pre-intervention and three post-intervention time periods to generate reliable results. For this analysis, collisions were accessed from the VDOT database from January 2010 through June 2016 on a monthly basis. This window of 42 pre-intervention and 36 post-intervention periods more than satisfices the design's minimum.

ITSA also requires the formation of data in a balanced set. Each time period must be uniform and consist of at least one entry of a dependent variable. The dependent variables were total statewide collisions and the subsets of serious injury and fatalities for moped operators occurring monthly. Each time period entry (i.e., month) also contained independent variables that the review of the literature recommended for inclusion. These included population estimates for the particular time period, non-moped-involved collisions, and unleaded regular gasoline fuel prices. Population figures for the months in between the official July intercensal estimates were interpolated with a linear growth rate as determined by the published estimates. Non-moped collisions for each month were extracted from the VDOT dataset and were included to serve as a controlled event series undisturbed by the moped legislation. Finally, average gasoline prices for each month were included to account for the theoretical highly elastic relationship between increased moped use (and exposure) and increasing fuel prices. Sharply climbing transportation fuel prices have been anecdotally tied to the increased level of mopeds on the state's roads, likely due to the class of vehicle's high fuel economy.

Visual analysis of moped collision and serious injury frequencies displayed strong seasonality. During the May through September months across the study period, collisions jumped approximately 400% compared with non-warm-weather months. Although VA's automobile collisions also display seasonality, summertime collision increases were only 20% to 30% during the same period. Controlling for the effect of the summer months will increase the internal validity of the study. The breadth of this study does not allow for the exploration of the source of moped seasonality, though it may be likely related to the physically exposed nature of the moped rider as compared with drivers and occupants within an enclosed vehicle. Thus, seasonality is represented through the creation of a dummy variable for the months May, June, July, August, and September.

After controlling for seasonality, population, non-moped vehicle incidents, and fuel prices, the format of the dependent variables must match the design's parameters. Calculating crash rates is typically accomplished through the comprehensive analysis of crash outcomes against vehicular exposure (such as number of licensed drivers or VMT). However, calculating crash rates for moped operators is problematic because both expected measures of exposure—licensed operators and VMT—are not available during the study period. An alternative path is to create a dependent variable that is solely the statewide crash frequency during each month. This approach carries difficulties, too, given the potential for crash frequencies to demonstrate regression to the mean if the study period is not long enough. The study period encompasses more than three years prior to the intervention and three years post, which is a time frame that can reduce the influence of this phenomenon (*15*). Nevertheless, a crude rate is included in the dependent variables as a share of moped collisions per that state's monthly population. A rate based upon the population overall draws close to the actual potential population of moped operators given the limited moped-use requirement of being over the age of 16 (no license or testing required). Including and testing both these formats—frequency and rate—will allow for a robustness check on the policy's effectiveness in reducing collisions, serious injuries, and fatalities.

An additional time series tool, the comparative interrupted time series analysis, was initially considered for the study's methodology. The State of North Carolina—which did not have a moped registration requirement during VA's moped policy enactment—was intended to serve as a control given its similar topography, climate, population, and energy prices, among other factors. However, the state's significant variation in North Carolina's vehicular collisions, presence of an existing moped helmet policy, and brief pre- and post-evaluation periods made sound evaluation problematic.

Logistic Regression of Collision Outcomes during Policy Phases

The final component of the research design is the application of a logistic regression to the individual characteristics of VA's collision-involved moped operators. This second analysis offers another opportunity for a robustness check of the effectiveness of the two-phased policy in reducing negative rider outcomes. It also provides an opportunity to examine the demographics and socioeconomic characteristics of riders from of an under-studied motorcycle subclass. The crash dataset is well suited for logistic regression analysis, as the odds ratio (OR) of serious crash injury and fatality can be computed across a series of independent variables, such as the implementation of a policy, demographics, and crash conditions.

As noted in the literature review, crash predictors range from personal characteristics to roadway conditions. Independent variables that serve as controls include the age, gender, alcohol consumption status, average weekly wage data for the county where the crash occurred, and helmet use of the moped operator him- or herself. Alcohol consumption was noted in circumstances in which alcohol was recorded as having been consumed prior to the collision; although the degree to which alcohol was a factor in the collision is necessary for legal culpability, determination at the crash site is subjective. The very consumption prior to operating the vehicle has been identified as a risk-taking behavior and crash predictor. Average weekly wage data for the county and quarter where and when the crash occurred was selected based on the predicted inverse relationship between economic activity and moped use.

Roadway and environmental conditions that also serve as independent variables include seasonality (coded similarly as ITSA), rural/urban designation, and the number of lanes where the crash occurred. The rural/urban designation was determined directly from the crash report. Urban or rural character may address issues such as crash exposure (i.e., more automobile exposure and crash risk in population-heavy urban areas) and travel speeds (i.e., assumed higher in rural areas), among others. Finally, the inclusion of number of lanes may function as an indicator of travel speeds and crash exposure.

The effects of the moped helmet and registration policies are included in a different format than that of ITSA. Collisions occurring during the pre-intervention Phase I and Phase II periods are coded accordingly. To avoid multicollinearity, Phase I and Phase II implementation statuses are included as separate dummy variables.

Dependent variables for the logistic regression analysis address only serious injury and death. The dataset represents the population of all collisions, thereby eliminating the possibility of evaluating the risk of being involved in a collision. Dummy variables were created for serious injury and fatality.

Demographic data completeness is an issue of consideration for analyzing the VDOT crash dataset. Though high-level collision information is available for all crashes (such as injury status of the moped operator), variables such as age, gender, number of lanes, and so forth, contain null or "not provided" values. For entries with null or "not provided" values, null values were blanketly applied. The resulting sample size for the logistic regression analysis is therefore reliant on the count of complete collision entries given the included dependent variables. Samples sizes are disclosed at the beginning of the logistic regression outputs.

Results

Descriptive Statistics

The population of recorded VA moped collisions between January 2010 and July 2016 was 3,207 collisions. Of these collisions, 571 (18%) occurred during the Phase I period, and 988 (31%) during Phase II. Notably fewer collisions are within the Phase I period given that that period was only 12 months.

The literature on VRU road safety has identified age and gender as risk predictors for collisions and injury. Table 1 below depicts numerous descriptive statistics according to the recorded gender of the moped operator. Male moped operators tend to be older, had consumed alcohol prior to the collision, collide in more rural areas, reside in counties with lower average weekly wages, and are less likely to be wearing a helmet than their female counterparts. Furthermore, male collisions outnumber female collisions by nearly 4:1.

ITSA Results

ITSA summary statistics convey the range of moped over 78 periods (months). The mean collision value per month is 41, with a range of 8 to 76. The ITSA dataset addresses non-moped collisions to control for roadway conditions that would not be affected by the moped operator policy. The time series plot of the ITSA dataset communicates the range of collisions over the study period. Figure 1 below displays the cyclical seasonal pattern of moped collisions from 2010 to mid-2016; the vertical bars represent

Table 1. Virginia Moped Operator Demographics

	Female moped operators			Male moped operators			Gender unknown		
Variable	Obs	Mean	SD	Obs	Mean	SD	Obs	Mean	SD
Driver age	481	34.89	14.04	2,097	37.94	14.65	435	34.08	14.44
Alcohol consumption	441	0.09	0.28	1,937	0.19	0.39	418	0.17	0.38
Urban area	398	0.86	0.34	1,797	0.82	0.38	398	0.86	0.35
Helmet worn	415	0.94	0.23	1,914	0.93	0.26	392	0.90	0.30
Number of lanes	398	3.26	1.39	1,797	3.26	1.41	398	3.21	1.41
Average weekly county wage (by US$100)	493	9.00	2.80	2,183	8.79	2.70	525	9.05	2.62

Note: Obs = observations; SD = standard deviation.

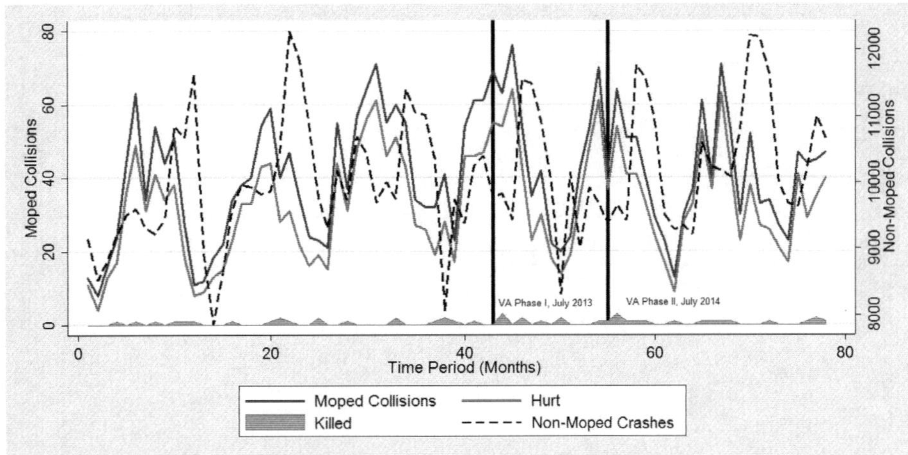

Figure 1. Virginia moped and motor vehicle collisions January 2010 through June 2016.

Table 2. Interrupted Time Series Stata Results for ITSA Model Runs

			Phase I			Phase II		
Model	Dependent variable	Variation	Coeff	0.1 α	0.05 α	Coeff	0.1 α	0.05 α
ITSA	Collision frequency	N-W	NA	NA	NA	NA	NA	NA
	Collision frequency	P-W	NA	NA	NA	NA	NA	NA
	Hurt frequency	N-W	NA	NA	NA	NA	NA	NA
	Hurt frequency	P-W	NA	NA	NA	NA	NA	NA
	Killed frequency	N-W	NA	NA	NA	NA	NA	NA
	Killed frequency	P-W	NA	NA	NA	NA	NA	NA
	Collision rate	N-W	−1.86E-07	*	NA	NA	NA	NA
	Collision rate	P-W	−1.96E-07	*	NA	NA	NA	NA
	Hurt rate	N-W	NA	NA	NA	NA	NA	NA
	Hurt rate	P-W	NA	NA	NA	NA	NA	NA
	Killed rate	N-W	NA	NA	NA	NA	NA	NA
	Killed rate	P-W	NA	NA	NA	NA	NA	NA

Note: Coeff = the model coeffecient; NA = results that were not statistically significant.
*Significance at <0.1 α. **Significance at <0.05 α. ***Significance at <0.001 α.

the Phase I and Phase II policy interventions at months 43 and 55. Non-moped roadway collisions are represented by the dashed black line on the second vertical axis.

The results of the series of ITSA are depicted below in Table 2. The dependent variable was modified to address both frequency and rate of collisions, injuries, and fatalities. Independent variables for the ITSA model included non-moped collisions, adjusted fuel prices, interpolated monthly population, and season (dummy variable for presence of warm weather). When the dependent variable was rate-based, interpolated population was dropped from the set of independent variables, given that the rate was moped outcomes standardized by population. Two variations of the ITSA command were used: Newey–West (N-W) and Prais–Winsten-Cochrane-Orcutt (P-W). Though both are variations of a time series ordinary least-squares regression model, the N-W produces N-W standard errors to handle autocorrelation in addition to possible heteroskedasticity, whereas the P-W uses the generalized least-squares method to estimate the parameters in a linear regression model in which the errors are assumed to follow a first-order autoregressive process (*14*). Diagnostic checks were preformed after each P-W and N-W model result for autocorrelation and appropriateness of lag periods through the Stata® "actest9" and Durbin–Watson commands. Both model variation results are presented for comparison purposes.

Table 3. Logistic Regression on Crash Predictors for Serious Injury Status of Virginia Moped Operators

Logistic regression

Number of observations = 1,662
LR $chi^2(10)$ = 41.11
Prob > chi^2 = 0
Pseudo R^2 = 0.030
Log likelihood = –691.79754

Independent variable	Odds ratio	Standard error	z	P>z
Phase I	1.17	0.23	0.78	0.44
Phase II	1.12	0.17	0.72	0.47
Wage (by US$100)	0.96	0.02	−1.71	0.09*
Summer	1.58	0.22	3.29	0.00***
Helmeted	0.53	0.20	−1.68	0.09*
Number of lanes	1.05	0.05	1.02	0.31
Urban area	0.56	0.14	−2.36	0.02**
Male	1.11	0.19	0.61	0.54
Age (by 10 years)	1.03	0.05	0.64	0.52
Alcohol consumption	1.60	0.34	2.19	0.03**
Gas price (adjusted 2016 USD)	1.10	0.09	1.10	0.27
_cons	11.75	6.16	4.70	0.00

Note: z = the Z score; P > z = the p-value.
*Significance at <0.1 α. ** Significance at <0.05 α. ***Significance at <0.001 α.

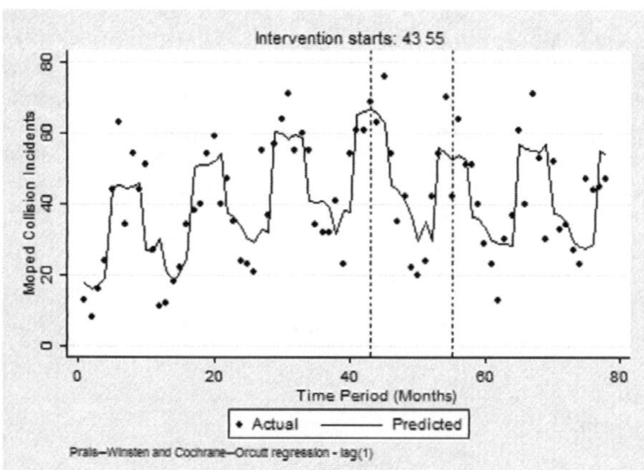

Figure 2. ITSA result for Virginia crash frequency, PW variation.

Results from ITSA show that monthly collision rates fell below predicted levels after the Phase I implementation but were not reduced further after Phase II. The coefficients for both the N-W and P-W models are significant at the 0.1 alpha level and translate to a reduction of 1.5 collisions per month following the implementation of Phase I. Statistically significant reductions were not observed for injury and fatality outcomes after either Phase I or Phase II.

The graphical representation in Figure 2 displays the actual and predicted collision frequencies for the ITSA model. Phased interventions are noted by the vertical lines.

Logistic Regression Moped Operator Outcomes Results

The logistic regression results for serious injury and fatality outcomes of moped operators do not show statistically significant ORs for Phase I and Phase II (see tables 3 and 4 below). Independent variables that were significant for serious injury included wage (4% increased risk for a reduction in US$100 weekly wages), summertime moped operation (58% increased risk as compared with non-summer), use of a helmet (47% reduction in risk), and crash in a rural location (44% increased risk as compared with urban areas). Alcohol consumption also increased the odds of serious injury by 60%. The only statistically significant predictor for fatalities was age, with every 10 years of age increasing fatality risk by 66%.

Discussion

The results of the two-part methodology do not provide a clear, coherent, and cohesive narrative for which to describe the effectiveness of VA's SB 1038 in generally reducing negative collision outcomes. Each design component—although intended to provide a robustness check on the other—used different methods and

Table 4. Logistic Regression on Crash Predictors for Fatality Status of Virginia Moped Operators

Logistic regression

Number of observations = 1,561
LR $chi^2(10)$ = 15.29
Prob > $chi^2 0$ = 0
Pseudo R^2 = 0.086
Log likelihood = –81.988038

Independent variable	Odds ratio	Standard error	z	P>z
Phase I	2.40	1.55	1.35	0.18
Phase II	2.12	1.82	0.87	0.38
Wage (by US$100)	0.90	0.10	–0.96	0.34
Summer	0.70	0.36	–0.69	0.49
Helmeted	1.00	(omitted)		
Number of lanes	0.84	0.19	–0.81	0.42
Urban area	1.56	1.11	0.62	0.53
Male	0.71	0.48	–0.51	0.61
Age (by 10 years)	1.65	0.30	2.79	0.01*
Alcohol consumption	1.87	1.05	1.12	0.26
Gas price (adjusted 2016 USD)	1.35	0.44	0.92	0.36
Constant	0.00	0.00	–3.18	0.00

Note: z = the *Z* score; *P* > z = the *p*-value.
*Significance at <0.1 α. **Significance at <0.05 α. ***Significance at <0.001 α.

selections of the overall source datasets, thereby requiring cautiousness in interpreting the results as a whole.

Results from ITSA indicate that Phase I reduced collisions for the months following its implementation, but that Phase II had no similar effect. Safety policy literature suggests that the direction of this effect may be in part due to the ID and helmet requirement's in deterring unsafe road practices through formalizing the mode and internalizing safety measures. Moped operators were expected to behave like other motor vehicles, and may have adjusted their behaviors. Another theory behind the reduction is the sample attenuation caused by the requirements of Phase I: risk-prone individuals may have removed themselves from the pool of moped operators. Surprisingly, the Phase II requirement of moped registration had no effect on reducing collisions or other negative outcomes. From the aforementioned logic, Phase II should have presented another obstacle to risk-prone individuals for operating a moped, thereby reducing collisions. It may be that Phase II—and policies like it—may not have a safety effect in additional to their basic record-keeping functions.

The logistic regression of collision outcomes supports ITSA's lack of effect of the two-phased policy on serious injury and fatal moped operator outcomes. Neither serious injuries nor fatalities had statistically significant ORs, meaning that being in those phases had no effect on the risk of being seriously hurt or injured. Unfortunately, the logistic regression component cannot be used to substantiate the results of the ITSA evaluation of Phase I; the collision dataset is a population of all reported collision events, and the statistical tool is not designed for a time series analysis. A potential issue with the comparison of the logistic regression's outcome to that of ITSA is the reduced sample size for the former. Given the demand of numerous demographic crash predictors (seven in addition to the seasonal and policy phase designations) in the logistic regression and the incompleteness of police report records, the sample size of complete person-level data entries was reduced from 3,207 to ˜1,600 collisions. There is nothing to indicate a systematic pattern of incompleteness for the collision records, nor is there support for random incompleteness. The effect of this reduction in sample size is unknown.

This study attempted to evaluate the effectiveness of VA's two-phased moped safety regulations despite several reporting and information obstacles. These obstacles included a lack of traditional measures of rider exposure (VMT and license or registration numbers) and an absence of evaluating crash outcomes over a compressed timeline that exhibited strong seasonality. Attempts were made to address these issues by standardizing crash frequency by population and coding for summer months. For future analyses of moped safety, the collection of registration data and long-term collision record keeping will largely eliminate these methodological problems. However, for an assessment of the policy intervention in the years leading up to and immediately following the measure's introduction, additional time series tools might be explored, such as an autoregressive moving average design. Furthermore, emergency room information could be combined with the police collision dataset to provide a more complete picture of moped collisions for policy evaluation purposes.

Conclusion

The implementation of VA's moped operator policy SB 1038 appears to have reduced the moped collision rate from mid-2013 to mid-2016. However, the policy did not result in a reduction of the frequency and risk of serious injury and fatality for moped operators during the same period. Though one of the policy's goals can be seen as successful through its downward pressure on collisions, the question remains as to why neither injuries nor fatalities were reduced. One potential explanation includes sample attenuation; moderate risk-taking individuals removed themselves from the pool of moped riders due to new regulation, thereby reducing collisions, but high risk-taking individuals were not deterred and continued contributing to serious and fatal moped outcomes. An alternative interpretation of the collision and injury/fatality rate divergence is that the moped policy did not change the harmful physics of moped collisions, unlike other vehicular interventions such as automobile seat belts. Yet another explanation includes the measurement of moped collision exposure. Without knowledge of the registered moped operator population or their VMT, the general population functioned as an exposure substitute. Future evaluation of the policy will likely benefit from enhanced moped population information, although any comparison of pre-policy outcomes will continue to rely on creative measures of exposure.

These results may be valuable to policy makers in other states without comparable moped operator laws. As the study demonstrated, collision rates fell after the initial introduction of the two-phased policy and remained lower than predicted levels while controlling for gasoline prices, seasonal variation, and roadway conditions. In 2013, VA experienced an estimated economic cost of US$747 million from property motor vehicle crashes alone, or approximately US$6,134 per crash incident (Virginia Highway Safety Office, n.d.). Even a minor reduction in non-serious injury collisions—as seen after SB 1038's implementation—would likely decrease costs associated with property damage, let alone costs associated with minor injuries. For the remaining states without moped operator requirements, there is an opportunity to both improve public safety and reduce economic losses through the adoption of basic moped safety and operator requirements. As the range of transportation options continues to grow, the public safety and transportation research community should further explore how overlooked modes, such as mopeds, and their operators respond to policy interventions. Doing so will allow policy makers to create safer environments for their citizens to more safely experience, traverse, and engage their communities.

Acknowledgments

This master's project would not have been possible without the direction, guidance, support, and feedback from numerous academic mentors and public agencies across the mid-Atlantic. The author acknowledges the support from the Dwight D. Eisenhower Transportation Fellowship that provided the financial means to complete the research. It is the author's hope that the research into moped users will better inform state-level public policies for these often misunderstood road users. The author expresses sincere thanks to the academic mentors during this year-long project. Daniel Rodriguez of the Universtiy of North Carolina at Chapel Hill (UNC) and Lindsay M. Braun of the University of Illinois at Urbana–Champaign shared many hours during the project's conceptual stage that shaped the author's approach. Noreen McDonald and William Lester of UNC provided feedback, comments, and reassurance during the analysis and presentation stages. UNC's City and Regional Planning Department was instrumental in the project's culmination.

Author Contributions

The author confirms contribution to the paper as follows: study conception and design: JS; data collection: JS; analysis and interpretation of results: JS; draft manuscript preparation: JS. The author reviewed the results and approved the final version of the manuscript.

References

1. National Highway Traffic Safety Administration. *Traffic Safety Facts: Motorcycles - 2013 Data.* National Highway Traffic Safety Administration, Washington, D.C., 2015, pp. 1–8. http://dx.doi.org/10.1016/j.annemergmed.2013.12.004.
2. U.S. Department of Transportation. *Fatality Analysis Reporting System [Data file].* 2016. http://www.nhtsa.gov/FARS.
3. Blackman, R., and N. Haworth. Comparison of Moped, Scooter and Motorcycle Crashes: Implications for Rider Training and Education. Presented at International Motorcycle Safety Conference, Orlando, FL, 2013. https://msf-usa.org/downloads/imsc2013/Oct17_Session1-Blackman-Comparison_of_Moped_Scooter_and_Motorcycle_Crashes_SLIDES.pdf. Accessed January 11, 2017.
4. Braun, B. M., L. M. Gries, A. N. Hildreth, and A. N. Miller. Moped Collisions among Patients with Revoked Drivers' Licenses Are a Significant Public Health Problem: A Retrospective Cohort Study. *American Surgeon*, Vol. 80, No. 8, 2014, pp. 792–795.
5. Miggins, M., L. Lottenberg, H. Liu, L. Moldawer, P. Efron, and D. Ang. Mopeds and Scooters: Crash Outcomes in a High Traffic State. *The Journal of Trauma*, Vol.

71, No. 1, 2011, pp. 217–222. http://doi.org/10.1097/TA.0b013e318208f874.

6. Haworth, N., and A. Nielson. Motor Scooters and Mopeds: Are Increasing Sales Translating into Increasing Crashes? *Transportation Research Record: Journal of the Transportation Research Board*, 2008. 2074: 69–76.
7. Lin, M.-R., and J. F. Kraus. A Review of Risk Factors and Patterns of Motorcycle Injuries. *Accident Analysis & Prevention*, Vol. 41, No. 4, 2009, pp. 710–722. http://doi.org/10.1016/j.aap.2009.03.010.
8. Staton, C., J. Vissoci, E. Gong, N. Toomey, R. Wafula, J. Abdelgadir, and M. Hocker. Road Traffic Injury Prevention Initiatives: A Systematic Review and Metasummary of Effectiveness in Low and Middle Income Countries. *PLoS One*, Vol. 11, No. 2, 2016, p. e0150150. http://doi.org/10.1371/journal.pone.0144971.
9. Law, T. H., R. S. Umar, S. Zulkaurnain, and S. Kulanthayan. Impact of the Effect of Economic Crisis and the Targeted Motorcycle Safety Programme on Motorcycle-Related Accidents, Injuries and Fatalities in Malaysia. *International Journal of Injury Control and Safety Promotion*, Vol. 12, No. 1, 2005, pp. 9–21. https://doi.org/10.1080/17457300512331339166.
10. Masten, S. V., and R. D. Foss. Long-Term Effect of the North Carolina Graduated Driver Licensing System on Licensed Driver Crash Incidence: A 5-Year Survival Analysis. *Accident Analysis and Prevention*, Vol. 42, No. 6, 2010, pp. 1647–1652. http://doi.org/10.1016/j.aap.2010.04.002.
11. Billheimer, J. W. California Motorcyclist Safety Program, Program Effectiveness: Accident Evaluation. *Transportation Research Record: Journal of the Transportation Research Board*, 1998. 1640: 100–109.
12. Kardamanidis, K., A. Martiniuk, R. Q. Ivers, M. R. Stevenson, and K. Thistlethwaite. Motorcycle Rider Training for the Prevention of Road Traffic Crashes. *Cochrane Database of Systematic Reviews*, n.d. http://doi.org/10.1002/14651858.CD005240.pub2.
13. Virginia Department of Motor Vehicles. *Non-Convential Vehicles Study – 2012 Report*. 2012. https://rga.lis.virginia.gov/Published/2012/RD380/PDF.
14. Linden, A. Conducting Interrupted Time-Series Analysis for Single- and Multiple-Group Comparisons. *The Stata Journal*, Vol. 15, No. 2, 2015, pp. 480–500. http://www.stata-journal.com/article.html?article=st0389.
15. Federal Highway Administration. *Appliying Safety Data and Analysis to Performance-Based Transportation Planning*. 2015. https://safety.fhwa.dot.gov/tsp/fhwasa15089/data_anl.pdf.

The Standing Committee on Motorcycles and Mopeds (ANF30) peer-reviewed this paper (18-05390).

Article

TRR
JOURNAL OF THE TRANSPORTATION RESEARCH BOARD

Transportation Research Record
2018, Vol. 2672(34) 106–121

DOI: 10.1177/0361198118798177
journals.sagepub.com/home/trr
SAGE

Understanding Emerging Motorcyclist Segments in Crashes using Florida Crash Data and Statewide Survey

Chanyoung Lee[1], Behzad Karimi[1], Siwon Jang[1], and Victoria Salow[2]

Abstract

The United States experienced a continued sharp increase in motorcycle fatalities between 1997 and 2008, with a 9% average annual increase, which resulted in a rapid doubling of motorcycle fatalities within a decade. After a major decline in both the number of fatalities and the fatality rate between 2008 and 2009, motorcycle fatalities and fatality rates have been fluctuating. It was discovered that the demographics of motorcyclists involved in fatal motorcycle crashes have changed because of an increase in the number of motorcyclists under age 30 and over age 50 during the past 10 years. As a result, motorcyclists in the United States can be clustered into three distinct segments by age in crash experiences. This study used motorcycle crash data in Florida that observed similar demographic changes and explored characteristics of the three segments. Descriptive analyses including Chi-square tests showed that each segment had different crash outcomes and different levels of exposure to common risk factors. Crash injury models were developed to understand variables that increase the level of injury severity in each segment. A statewide survey was completed to explore the sociodemographic characteristics of the three motorcyclist segments in Florida, which often are not available through crash data. Overall, the three motorcycle segments identified in this study have unique riding characteristics and crash outcomes. This is essential information for developing and managing motorcycle safety programs in an effective and efficient manner.

The United States experienced a continued sharp increase in motorcycle fatalities between 1997 and 2008, with a 9% average annual increase, which resulted in a rapid doubling of motorcycle fatalities within a decade. Although the United States also observed a somewhat similar increasing trend in motorcycle registrations during the same period, the rate of increase in motorcycle fatalities far exceeded the rate of increase in motorcycle registrations. As a result, motorcycle fatalities per 100,000 registered motorcycles increased from 55 fatalities in 1996 to 69 fatalities in 2008. As shown in Figure 1, a major decline in motorcycle fatalities was observed in both the number of fatalities and the fatality rate between 2008 and 2009, which also coincided with a recession in the US economy. Since the decline, motorcycle fatalities and fatality rates in the United States have been fluctuating, but it is notable that motorcycles have consistently represented 14% of motor-vehicle fatalities for the past 5 years. During the same period, motorcycles represented 3% of all annual motor-vehicle registrations and accounted for less than 1% of yearly vehicle miles traveled (VMT). (*1*)

Although motorcycles remain one of the focus areas in the Strategic Highway Safety Plan (SHSP) in many states, it is challenging to promote motorcycle safety in an effective and efficient manner for various reasons, including limited resources. First, it is well-known that motorcycles are vehicles with significant inherited risk compared to passenger cars. Motorcycles provide very limited protection to their occupant(s), and motorcyclists often are ejected from the vehicle in traffic crashes. Therefore, efforts to improve motorcycle safety in the United States often focus on improving motorcyclist skills, attitudes, and behaviors.

Second, motorcyclists represent a small portion of the motoring public in the United States; however, it is a group that has a well-developed and diverse subgroup culture. Therefore, it is important to comprehend various motorcycling population segments that are highly associated with motorcycle type and unique values to develop

[1]Center for Urban Transportation Research, University of South Florida, Tampa, FL
[2]Injury Prevention & Research Center, Ann & Robert H. Lurie Children's Hospital of Chicago, Chicago, IL

Corresponding Author:
Address correspondence to Chanyoung Lee: leec@cutr.usf.edu

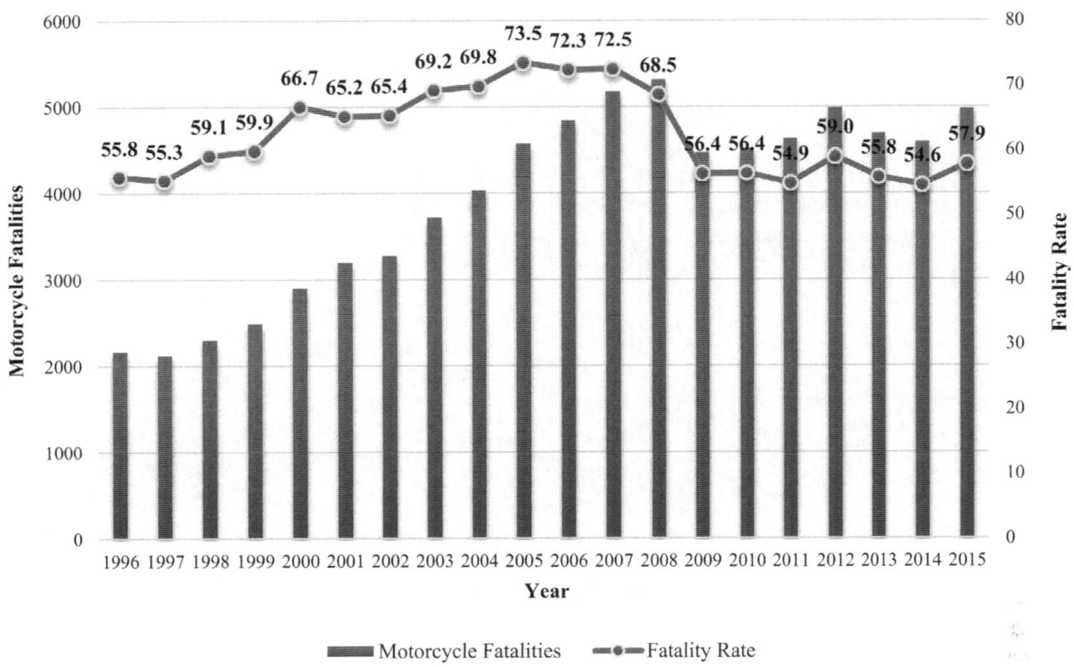

Figure 1. Motorcycle fatalities and fatality rates (per 100,000 registered vehicles) in the United States.

effective behavioral safety intervention programs for motorcyclists. Third, the demographics of motorcyclists who have crashed in the United States are changing over time, which may require a timely update of messages and intervention approaches. Figure 2 shows two sets of three years of motorcycle fatalities in the United States by age for 2003–2005 and 2013–2015. Although there was a slight change in the total number of motorcycle fatalities between 2005 and 2015, it is interesting to note that the changing demographics of motorcycle fatalities has been significant during the past 10 years. It is somewhat expected to see two peaks in motorcycle fatalities, which is different from a typical single peak in young drivers in passenger car fatalities. However, the figure shows that the two peaks have become much more obvious, with a significant increase in fatalities of motorcyclists age 50 and over.

This study aimed to understand the characteristics of motorcycle crashes of three age groups (under 30, 30–49, 50 or older) and to extend the knowledge about motorcycle riding and lifestyles in each group. However, it was difficult to obtain nationwide motorcycle crash data that include non-fatal crashes, and conducting a nationwide survey with a representative sample was not feasible because of limited resources. At the same time, Florida experienced a very similar demographic change in motorcycle crashes. Figure 3 shows the demographic changes in seriously or fatally injured motorcyclists in Florida during the same period. Florida represents about 10% of US motorcycle registrations and US motorcycle fatalities. In Florida in 2015, there were 584 motorcycle crash fatalities and 9,045 injury-related motorcycle crashes. Therefore, the scope of study was limited to the state of Florida.

As of July 1, 2015, Florida had 1,185,787 drivers with motorcycle endorsements and 601,191 registered motorcycles, representing about 3.4% of Florida-registered vehicles (not including mobile homes and vessels). Although it is difficult to measure the actual usage of motorcycles because of limitations in counting them with existing traffic detectors, it is estimated that motorcycles represent less than 1% of traffic on Florida's roadways, based on Florida vehicle classification information. However, motorcycle fatalities represented about 20% of Florida traffic fatalities in 2015.

Literature Review

During the past 10 years, numerous reports have warned of the alarming increase in motorcycle fatalities in the United States (*1–3*). Many efforts have been made to investigate motorcycle crashes and explore potential countermeasures. The following summarizes previous motorcycle safety studies that are relevant to motorcyclist characteristics by age.

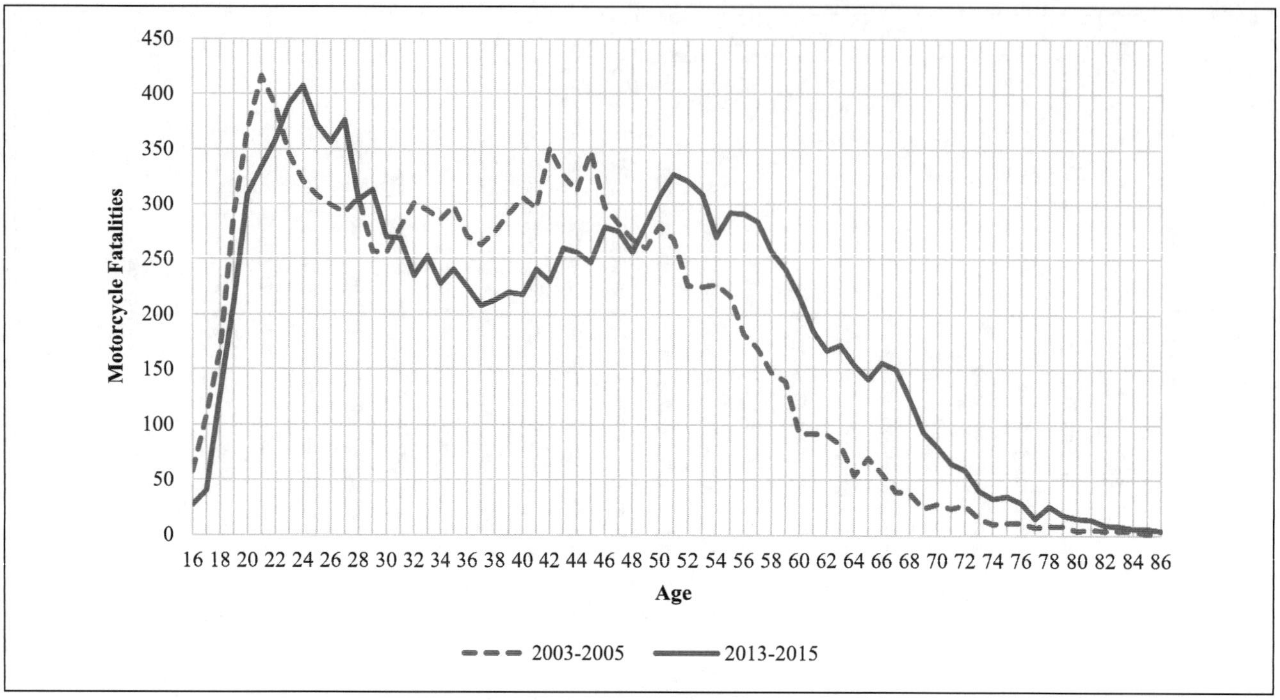

Figure 2. Changing demographics of motorcycle fatalities in the United States.

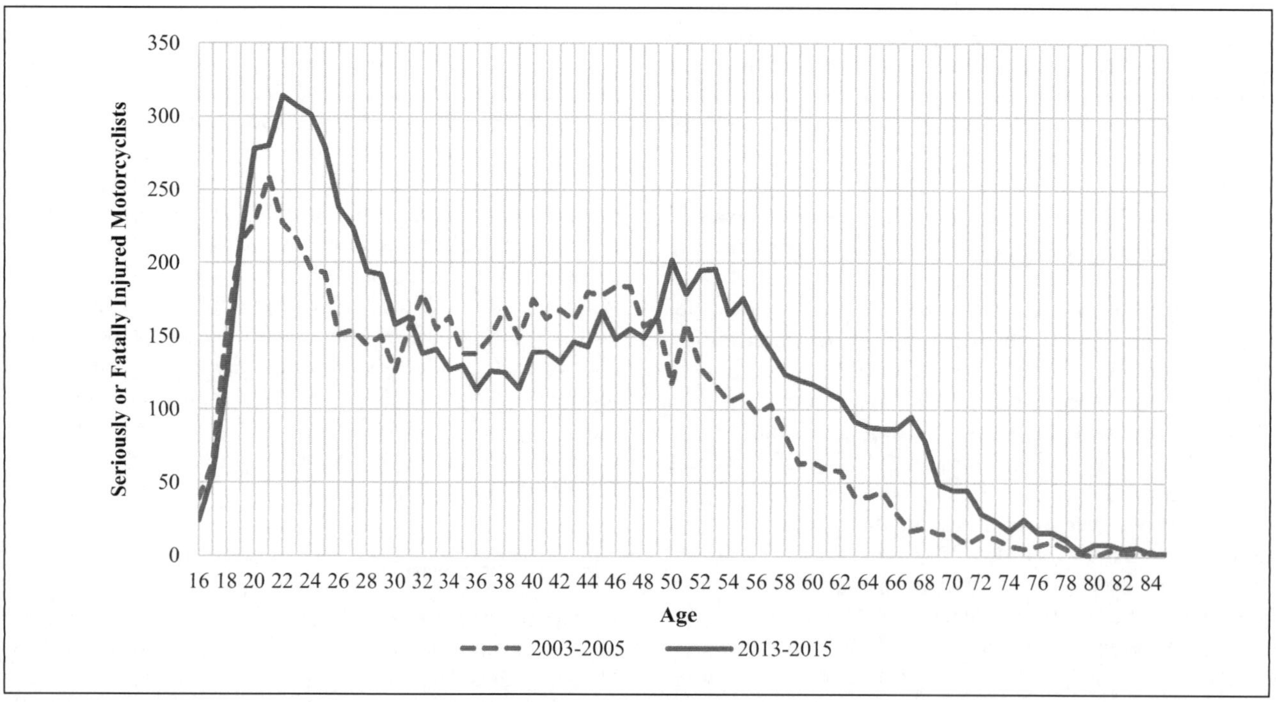

Figure 3. Changing demographics of seriously or fatally injured motorcyclists in Florida.

Riding Experience

In the United States, more than half of active motorcyclists age 50 or older returned to riding after taking a continuous break of at least 2 years (average length 11 years), and 20% returned after taking a break of 20 years or longer (*4*). Increasing age is positively associated with increasing annual VMT in the United States, such that motorcyclists ages 18–29 and 50–59 ride an average of 4,268 and 6,437 miles per year, respectively (*4*). Nearly half (46%) of all fatally injured motorcyclists aged 20 or younger are improperly licensed compared to only 18% and 9% of motorcyclists ages 40–49 and 50 or older, respectively, in the United States (*5*).

Motorcycle Choice

Larger motorcycles are associated with higher fatality rates than smaller motorcycles (*6–8*), and older motorcyclists own larger motorcycles than younger motorcyclists (*8–10*). According to Teoh and Campbell (*8*), the average age of touring riders is 51 and the average age of sport-type motorcycle riders is 27. Only 2% of touring and 12% of cruiser or standard motorcycle riders were younger than 30 in the study. In the United States, nearly three-quarters (72%) of all fatally injured motorcyclists who had been riding bikes with engine sizes 1,001–1,500 cc at the time of the crash were age 40 or older (*5*). Further, among motorcyclists age 40 or older, motorcyclists on bikes with an engine capacity of 1,000 cc or larger were four times more likely to be killed in a crash than riders on bikes with smaller engine capacities (*11*).

Purpose of Riding

Younger motorcyclists are more likely to ride for commuting purposes and less likely to ride for recreational purposes than are older motorcyclists (*4, 10*). Lee et al. (*12*) found that riders of all ages surveyed in Florida were most likely to report riding for "the thrill of riding and the freedom of the open road," but the second most-common reason for riding among riders ages 45–64 was "I want to ride with friends"; "inexpensive means of transportation" was cited by riders ages 18–44.

Speeding

Older motorcyclists are significantly less likely to have positive attitudes toward risk-taking than younger motorcyclists, and positive attitudes toward risky behaviors are the greatest predictor of behaving in a risky manner (*9*). In the United States, more than half of all motorcycle fatalities among riders ages 20–29 are attributed to speeding compared to approximately one-quarter of fatalities among riders age 40 or older (*5*), and riders age 45 or older are only one-third as likely to be traveling 60 mph or faster at the time of a crash compared to younger riders (*13*).

Helmet Use

Younger motorcyclists (ages 18–29) are more likely to believe that helmets keep riders safe and to favor universal helmet legislation than are riders of any other age group (*4*). Younger motorcyclists also appear to be more likely to wear helmets than middle-age motorcyclists. Although data on helmet use by age are somewhat inconsistent, data from both a national survey of motorcyclists in the United States and the Fatality Analysis Reporting System (FARS) indicate that helmet use and age have a curvilinear relationship, such that usage rates are highest among the youngest and oldest motorcyclists and lowest among middle-age motorcyclists (*4, 5*).

Alcohol Use

Studies that dichotomized motorcyclists as older (i.e., age 40 and older) or younger (i.e., age 39 and younger) found no difference in alcohol-related crash involvement by age (*11, 14*). Studies that examine differences in alcohol-related crash involvement between clustered age groups (e.g., ages younger than 20, 20–29, 30–39, 40–49, 50–59, 60+), however, reveal that middle-age riders (ages 30–59) account for a disproportionately large share of crashes, with riders ages 40–49 and 30–39 at greatest and second-greatest risk, respectively (*5, 15–18*).

Crash Outcomes

Older age is a significant predictor of fatality in the event of a crash (*13, 19*), such that fatality rates among motorcyclists consistently increase by 1.1–1.5% with each one-year increase in age (*20, 21*), and the risk of serious injury increases by 4.2% with each one-year increase in age (*22*). Older motorcyclists are significantly more likely to sustain a severe injury than a minor injury in the event of a crash (*23–25*), and motorcyclists ages 55–65 are overrepresented among serious injury crashes (*26*). Crash deaths are significantly more likely among riders under the age of 40 than among riders age 40 or older, and riders under the age of 29 sustain the most serious motorcycle damage in the event of a crash (*11, 27*).

Motorcycle Crashes in Three Age Groups

Data from the Florida Department of Transportation's (Florida DOT) Crash Analysis Reporting System (CARS) for 2013–2015 were used to conduct a descriptive analysis of motorcycle crashes by age group. The three age groups consisted of motorcycle operators age

Table 1. Motorcycle Crashes in Florida by Injury Severity and Age Group

Injury severity	Group A (under age 30)	Group B (ages 30–49)	Group C (age 50 or older)	Total
None	14.9% (1,654)	16.9% (1,675)	16.8% (1,531)	16.1% (4,860)
Possible	19.9% (2,209)	19.9% (1,970)	19.1% (1,745)	19.7% (5,924)
Non-incapacitating	39.8% (4,412)	37.2% (3,687)	35.8% (3,267)	37.7% (11,366)
Incapacitating	20.4% (2,262)	21.7% (2,152)	23.6% (2,157)	21.8% (6,571)
Fatal	4.9% (548)	4.3% (430)	4.8% (436)	4.7% (1,414)
Total	100% (11,085)	100% (9,914)	100% (9,136)	100% (30,135)

Note: Numbers in parentheses are frequency.

Table 2. Incapacitating/Fatal Injury Motorcycle Crashes by At-Fault and Age Group

Crash type	Group A (under age 30)	Group B (ages 30–49)	Group C (age 50 or older)	Total
Single-vehicle crash	32.7% (919)	38.6% (997)	39.1% (1,014)	36.7% (2,930)
Multi-vehicle crash (motorcycle at fault)	28.9% (813)	22.7% (588)	22.4% (580)	24.8% (1,979)
Multi-vehicle crash (other vehicle at fault)	38.4% (1,078)	38.7% (999)	38.5% (999)	38.5% (3,076)
Total	100% (2,810)	100% (2,582)	100% (2,583)	100% (7,985)

Note: Numbers in parentheses are frequency.

30 and under (Group A), ages 30–49 (Group B), and ages 50 or older (Group C), which were determined based on age thresholds of the increase/decrease in serious and fatal motorcycle crashes during the past 10 years in Florida. As shown in Table 1, the three age groups divided Florida motorcycle crash data into three comparably sized clusters. A Chi-square test shows that the relationship between injury severity and age group is significant, $\chi^2(\text{df} = 8, N = 30,135) = 68,668, p < .01$.

Table 2 divides motorcycle crashes with incapacitating or fatal injury into single- and multi-vehicle crashes. It appears that single-vehicle crashes represent about one-third of crashes in each group. A Chi-square test shows that the relationship between "crash type by at-fault" and age group is significant, $\chi^2(\text{df} = 4, N = 7,985) = 49.089, p < .001$. Groups B and C are more involved in single-vehicle crashes compared to Group A, and Group A had more motorcyclist at-fault crashes compared to the other two groups.

Three common risk factors in motorcycle crashes include lack of helmet use, riding under the influence of alcohol, and speeding. As shown in Table 3, Chi-square tests show that each group is associated with these risk factors at different levels. The proportion of motorcyclists with risk factors increased for all three groups in fatal crashes compared to all crashes. It seems that more motorcyclists in Group C wear no motorcycle helmet compared to the other two groups, and Group B has more impaired-riding crashes. It is notable that the proportion of Group A in fatal crashes that exceeded 20 mph over the posted speed limit is significantly higher than in the other two groups.

It is known that motorcycle type is highly associated with motorcyclist age, which is important for explaining riding behaviors and attitudes, as well as crash outcomes. However, many crash databases do not include motorcycle-type information, although it can be obtained by decoding the vehicle identification number (VIN) of motorcycles in police crash reports. In this study, 120 police crash reports for fatal motorcycle crashes were randomly selected and reviewed, including narrative and crash diagrams, to establish motorcycle crashes with motorcycle-type information. In summary, Group A crashes consisted mainly of sport bikes operated almost exclusively by males in the 20–29 age group. The principal operation error was excessive speed. In these crashes, sport bikes often exceeded the speed limit by a minimum of 20 mph (i.e., speed limit 55 mph, motorcycle traveling 75 mph) and reached $\geq$100 mph. With other drivers expecting a motorcycle to operate at a normal approach speed, this creates "failure to yield" charges for motorists violating the motorcyclist's right-of-way as a result of a crash and results in serious or fatal injury of the motorcyclist. This high-speed operation also leads to stop-sign and traffic-signal violations by motorcyclists.

Group B crashes are still dominated by sport bikes, but cruiser-style motorcycle representation has increased; this cluster includes males ages 30–50. The operators of sport motorcycles in this group reflect the same type of

Table 3. Common Risk Factors in Motorcycle Crashes (No Helmet, Drinking, Speeding)

Injury type	Category	Group A (under age 30)	Group B (ages 30–49)	Group C (age 50 or older)	Total	df	χ^2 (*p*-value)
Helmet use							
All crashes[a]	DOT-compliant motorcycle helmet	64.7%	48.5%	47.2%	54.1%	4	769.679 (<.001)
	Other helmet	1.9%	2.1%	2.7%	2.2%		
	No helmet	33.4%	49.4%	50.1%	43.7%		
	Total	100% (10,538)	100% (9,386)	100% (8,722)	100% (28,646)		
Fatal crashes	DOT-compliant motorcycle helmet	64.9%	44.5%	38.5%	50.5%	4	82.917 (<.001)
	Other helmet	1.5%	3.1%	6.3%	3.5%		
	No helmet	33.6%	52.4%	55.2%	46.0%		
	Total	100% (536)	100% (422)	100% (431)	100% (1,389)		
Suspected alcohol use							
All crashes[a]	No	91.9%	89.5%	91.3%	90.9%	4	72.528 (<.001)
	Yes	3.2%	5.4%	4.5%	4.3%		
	Unknown	4.9%	5.1%	4.3%	4.8%		
	Total	100% (11,099)	100% (9,924)	100% (9,146)	100% (30,169)		
Fatal crashes	No	49.4%	41.6%	53.7%	48.3%	4	18.041 (<.001)
	Yes	12.1%	18.1%	10.8%	13.5%		
	Unknown	38.6%	40.2%	35.6%	38.1%		
	Total	100% (547)	100% (430)	100% (436)	100% (1,413)		
Speeding (posted speed limit – estimated vehicle speed)							
All crashes[a]	Stopped	8.6%	12.0%	12.7%	11.0%	8	928.829 (<.001)
	At/under posted speed limit	71.1%	74.7%	81.0%	75.3%		
	1–10 mph above	8.5%	6.4%	4.0%	6.4%		
	11–20 mph above	5.8%	3.7%	1.5%	3.8%		
	21 mph or higher	6.0%	3.2%	0.8%	3.5%		
	Total	100% (10,912)	100% (9,814)	100% (9,074)	100% (29,800)		
Fatal crashes	Stopped	13.9%	9.7%	9.3%	13.9%	8	134.725 (<.001)
	At/under posted speed limit	38.6%	52.0%	72.4%	38.6%		
	1–10 mph above	8.2%	8.7%	7.9%	8.2%		
	11–20 mph above	15.6%	14.2%	5.1%	15.6%		
	21 mph or higher	23.7%	15.4%	5.3%	23.7%		
	Total	100% (539)	100% (423)	100% (431)	100% (1,393)		

Note: numbers in parentheses are frequency.
[a]All crashes also includes fatal crashes.

speed violations as their younger riding cohorts. Cruiser riders in this group, on the other hand, depart from the traveled portion of the roadway on curves or strike median curbs and traffic islands while under the influence. Group C consists mainly of cruiser and touring models, along with a fair number of scooters. Crashes in this group do not necessarily involve excessive speed; in addition to the common left turn across path (LTAP) scenario, they are often involved in sideswipe-type crashes with vehicles traveling in the same direction or striking the rear of a slowing or stopped vehicle. The number of intoxicated motorcyclists in this group is not greater than the other groups; however, there was a larger number of intoxicated "other vehicle" operators in the crashes.

Crash Injury Severity Modeling

The descriptive analysis of motorcycle crash data confirmed that each group has a different frequency of various motorcycle crash types and is associated with various risk factors at different levels. In other words, each group has a different magnitude in many variables that are associated with crash outcomes. To further investigate the effects of these risk factors, including other environmental factors on crash injury severity in each group, injury severity models were developed by using an ordered logit (OL) model.

Methodology

Motorcyclist injury severity was divided into five ordinal categories: none, possible, non-incapacitating, incapacitating, and fatal. The injury severity as an ordinal variable was modeled with an OL model. Let S_p be the discrete injury severity level of motorcyclist p. Then, S_p^*, the associated latent utility of injury severity S_p, is assumed to be:

$$S_p^* = \alpha' z_p + \omega_p \tag{1}$$

where z_p is the vector of observable independent covariates; α is the vector of coefficients to be estimated; and ω_p is the error term that follows standard logistic distribution with cumulative distribution function $G(.)$.

Let δ_k be cutoff points that divide S_p^* into K $(k = 1, 2, \ldots, K)$ discrete injury severity levels. Then, from Equation 1 it can be concluded that injury severity is equal to crash severity level k when

$$\delta_{k-1} < \alpha' z_p + \omega_p \leq \delta_k \tag{2}$$

where $\delta_{i,0} = -\infty$ and $\delta_{i,K} = +\infty$.

The probability that motorcyclist p is involved in a crash with severity level k can be written as:

$$\begin{aligned} P(S_p = k) &= P(\delta_{k-1} < \alpha' z_p + \omega_p < \delta_k) \\ &= P(\delta_{k-1} - \alpha' z_p < \omega_{pi} < \delta_k - \alpha' z_p) \\ &= G(\delta_k - \alpha' z_p) - G(\delta_{k-1} - \alpha' z_p) \end{aligned} \tag{3}$$

Finally, the parameters of the model are estimated by maximizing the following log-likelihood function:

$$\text{LL} = \sum_{p=1}^{N} \sum_{k=1}^{K} \text{q}_{p,k} {*} \log P(S_p = k) \tag{4}$$

where N is the total number of motorcyclists in the sample; and $\text{q}_{p,k} = 1$ when motorcyclist p was involved in a crash with severity level k, and 0 otherwise.

The parameters of the model are estimated with SAS econometrics software. The parameters to be estimated include α vector and cutoff points δ_k $(k = 1, 2, \ldots, K)$.

Data

Florida DOT's CARS data from 2013 to 2015 were used to estimate crash injury severity models. The CARS dataset includes all reported crashes with a fatality, an injury, or high property damages, and provides information about motorcyclist characteristics, roadway attributes, environmental factors, temporal and spatial attributes, and crash characteristics.

Estimation Results

Descriptive statistics of variables used in the models are presented in Table 4. The model is composed of the wide range of variables described previously. Table 5 shows the final estimation results. The injury severity level, as the dependent variable, was modeled as an ordinal variable using the OL model. The estimated parameters for each motorcycle type are presented under a separate column. The likelihood ratio test of the models showed that all estimated models are statistically significant at a level less than.0001. Positive (negative) coefficients increase (decrease) the level of injury severity. For example, riding a motorcycle under the influence of alcohol is a factor that increases the injury severity of motorcyclists in all three groups.

The factors that increase the level of injury severity across all age groups are: riding under the influence, using no safety equipment, curve segment of roadway, intersection, driveway access, and cloudy weather. Generally, as the speed limit increases, the risk of a severe injury increases, especially on highways with speed limits of 45 and 50 mph. Riding a motorcycle when impaired increases the likelihood of a more severe injury. Based on the estimated thresholds, a motorcyclist with a positive DUI (driving under the influence) and no safety

Table 4. Descriptive Statistics of Variables Used in Final Models

Variable	Definition	Group A (under age 30)		Group B (ages 30–49)		Group C (age 50 or older)	
		Mean	Std. dev.	Mean	Std. dev.	Mean	Std. dev.
Driver characteristics							
DUI status							
Positive DUI	1 if motorcyclist is under the influence of alcohol or drugs, 0 otherwise	0.02	0.12	0.03	0.17	0.03	0.16
Number of safety equipment							
0	1 if motorcyclist has no safety equipment, 0 otherwise	0.18	0.39	0.20	0.40	0.16	0.37
Roadway attributes							
Shoulder width (ft)							
0	1 if there is no shoulder, 0 otherwise	0.67	0.47	0.63	0.48	0.63	0.48
1–2	1 if shoulder width is between 1 ft and 2 ft, 0 otherwise	0.07	0.25	0.08	0.26	0.06	0.23
Speed limit (mph)							
≤30	1 if speed limit is 30 mph or less, 0 otherwise	0.28	0.45	0.25	0.43	0.24	0.43
35 and 40	1 if speed limit is 35/40 mph, 0 otherwise	0.28	0.45	0.27	0.44	0.25	0.44
45 and 50	1 if speed limit is 45/50 mph, 0 otherwise	0.32	0.47	0.33	0.47	0.34	0.48
≥55	1 if speed limit is 55 mph or more, 0 otherwise	0.12	0.33	0.15	0.35	0.16	0.37
Traffic way character							
Curve, level	1 if road segment is level curve, 0 otherwise	0.10	0.30	0.11	0.31	0.10	0.29
Curve, up and down grade	1 if road segment is upgrade/downgrade curve, 0 otherwise	0.03	0.16	0.03	0.16	0.02	0.15
Straight, up and down grade	1 if road segment is upgrade/downgrade straight, 0 otherwise	0.04	0.19	0.04	0.19	0.04	0.19
Traffic control							
Stop sign	1 if site location is an intersection with stop-sign control, 0 otherwise	0.10	0.30	0.10	0.30	0.10	0.30
Signal	1 if site location is an intersection with traffic-signal control, 0 otherwise	0.17	0.38	0.19	0.39	0.19	0.39
Environmental factors							
Light condition							
Daylight	1 if it is daylight, 0 otherwise	0.60	0.49	0.61	0.49	0.73	0.44
Dark without street lighting	1 if it is dark and without street lighting, 0 otherwise	0.08	0.27	0.09	0.28	0.07	0.26
Weather condition							
Cloudy	1 if it is cloudy, 0 otherwise	0.15	0.36	0.14	0.35	0.15	0.35
Road surface condition							
Wet and slippery	1 if road surface is wet and slippery, 0 otherwise	0.10	0.30	0.10	0.30	0.09	0.28
Temporal attributes							
Time of day							
1–4 a.m.	1 if crash occurs between 1 and 4 a.m., 0 otherwise	0.06	0.23	0.05	0.22	0.02	0.14
Weekday							
Sunday	1 if crash occurs on Sunday, 0 otherwise	0.16	0.37	0.18	0.39	0.20	0.40
Tuesday	1 if crash occurs on Tuesday, 0 otherwise	0.13	0.34	0.12	0.32	0.12	0.32
Thursday	1 if crash occurs on Thursday, 0 otherwise	0.14	0.35	0.14	0.35	0.12	0.33
Month							
April	1 if crash occurs in April, 0 otherwise	0.09	0.29	0.09	0.28	0.10	0.29
June	1 if crash occurs in June, 0 otherwise	0.08	0.27	0.07	0.26	0.06	0.25

(continued)

Table 4. (continued)

Variable	Definition	Group A (under age 30)		Group B (ages 30–49)		Group C (age 50 or older)	
		Mean	Std. dev.	Mean	Std. dev.	Mean	Std. dev.
July	1 if crash occurs in July, 0 otherwise	0.08	0.27	0.08	0.26	0.07	0.25
Spatial characteristics							
District							
District 3	1 if location of crash is in District 3, 0 otherwise	0.07	0.25	0.07	0.26	0.06	0.25
District 7	1 if location of crash is in District 7, 0 otherwise	0.18	0.38	0.17	0.38	0.18	0.39
County							
Bay	1 if location of crash is in Bay County, 0 otherwise	0.01	0.11	0.02	0.13	0.02	0.12
Broward	1 if location of crash is in Broward County, 0 otherwise	0.08	0.27	0.08	0.28	0.06	0.24
Citrus	1 if location of crash is in Citrus County, 0 otherwise	0.00	0.07	0.01	0.09	0.02	0.13
Escambia	1 if location of crash is in Escambia County, 0 otherwise	0.02	0.12	0.02	0.13	0.01	0.12
Duval	1 if location of crash is in Duval County, 0 otherwise	0.07	0.25	0.08	0.26	0.06	0.23
Lake	1 if location of crash is in Lake County, 0 otherwise	0.01	0.09	0.02	0.13	0.02	0.15
Manatee	1 if location of crash is in Manatee County, 0 otherwise	0.01	0.12	0.01	0.12	0.02	0.15
Marion	1 if location of crash is in Marion County, 0 otherwise	0.01	0.12	0.02	0.13	0.02	0.14
Miami-Dade	1 if location of crash is in Miami-Dade County, 0 otherwise	0.15	0.36	0.14	0.35	0.08	0.27
Monroe	1 if location of crash is in Monroe County, 0 otherwise	0.04	0.19	0.04	0.20	0.05	0.22
Okaloosa	1 if location of crash is in Okaloosa County, 0 otherwise	0.01	0.10	0.01	0.11	0.01	0.10
Palm Beach	1 if location of crash is in Palm Beach County, 0 otherwise	0.05	0.23	0.04	0.19	0.04	0.20
Pasco	1 if location of crash is in Pasco County, 0 otherwise	0.03	0.16	0.02	0.15	0.03	0.18
Pinellas	1 if location of crash is in Pinellas County, 0 otherwise	0.06	0.24	0.05	0.22	0.06	0.24
Site location							
Intersection	1 if site location is intersection, 0 otherwise	0.28	0.45	0.29	0.45	0.30	0.46
Driveway access	1 if site location is driveway access, 0 otherwise	0.04	0.21	0.04	0.20	0.05	0.21
Parking lot	1 if site location is parking lot, 0 otherwise	0.02	0.15	0.02	0.14	0.02	0.16
Motorcycle characteristics							
Body class							
Cruiser	1 if motorcycle body class is cruiser, 0 otherwise	0.19	0.39	0.39	0.49	0.49	0.50
Scooter	1 if motorcycle body class is scooter, 0 otherwise	0.10	0.29	0.09	0.29	0.10	0.30
Sport	1 if motorcycle body class is sport, 0 otherwise	0.15	0.36	0.09	0.28	0.02	0.14
Supersport	1 if motorcycle body class is supersport, 0 otherwise	0.42	0.49	0.20	0.40	0.02	0.13
Number of observations		10,344	8,845	7,594			

Table 5. Estimated Injury Severity Models for Age Group

	Group A (under age 30)			Group B (ages 30–49)			Group C (age 50 or older)		
Variable	Coeff.	*t*-stat	Odds ratio	Coeff.	*t*-stat	Odds ratio	Coeff.	*t*-stat	Odds ratio
Threshold parameters									
Injury severity									
None to possible	−2.89	−35.24	–	−3.55	−40.45	–	−3.18	−36.56	–
Possible to non-incapacitating	−0.95	−13.22	–	−1.40	−19.07	–	−1.00	−13.93	–
Non-incapacitating to incapacitating	0.84	11.67	–	0.31	4.32	–	0.63	8.84	–
Incapacitating to fatal	2.02	26.95	–	1.47	19.73	–	1.73	23.39	–
Driver characteristics									
DUI status									
Positive DUI	0.46	3.09	1.58	0.48	4.11	1.62	0.35	2.53	1.42
Number of safety equipment									
0	0.34	6.78	1.40	0.23	4.63	1.26	0.13	2.29	1.14
Roadway attributes									
Shoulder width (ft)									
0	0.14	2.96	1.16	0.16	3.51	1.17	–	–	–
1–2	−0.19	−2.75	0.83	–	–	–	−0.21	−3.12	0.81
Speed limit (mph)									
≤30	−0.74	−10.85	0.48	−0.43	−7.66	0.65	−0.49	−8.02	0.61
35 and 40	−0.52	−7.93	0.59	−0.19	−3.72	0.83	−0.24	−4.41	0.79
45 and 50	−0.30	−4.85	0.74	–	–	–	–	–	–
≥55	–	–	–	0.33	5.36	1.39	0.32	5.02	1.37
Traffic way character									
Curve-level	0.51	8.52	1.67	0.58	9.06	1.79	0.72	9.85	2.05
Curve, up and down grade	0.78	6.75	2.18	0.73	6.20	2.08	0.59	4.15	1.81
Straight, up and down grade	0.20	2.13	1.22	–	–	–	–	–	–
Traffic control									
Stop sign	–	–	–	–	–	–	0.22	3.00	1.25
Signal	−0.16	−3.16	0.85	−0.28	−5.37	0.75	−0.17	−2.87	0.85
Environmental factors									
Light condition									
Daylight	−0.19	−5.02	0.82	−0.14	−3.21	0.87	−0.18	−3.23	0.84
Dark without street lighting	–	–	–	0.26	3.43	1.30	0.27	2.85	1.31
Weather condition									
Cloudy	0.24	4.71	1.27	0.26	4.60	1.30	0.19	3.22	1.22
Road surface condition									
Wet and slippery	−0.32	−5.24	0.72	−0.25	−3.75	0.78	−0.20	−2.67	0.82
Temporal attributes									
Time of day									
1–4 a.m.	0.29	3.56	1.33	0.40	4.26	1.50	–	–	–
Weekday									
Sunday	0.13	2.45	1.14	0.21	4.02	1.24	–	–	–
Tuesday	–	–	–	–	–	–	−0.15	−2.27	0.86
Thursday	0.12	2.34	1.13	–	–	–	–	–	–
Month									
April	–	–	–	–	–	–	0.16	2.17	1.17
June	–	–	–	−0.17	−2.28	0.84	–	–	–
July	–	–	–	0.20	2.68	1.22	–	–	–
Spatial characteristics									
District									
District 3	–	–	–	−0.22	−2.74	0.80	–	–	–
District 7	0.28	4.91	1.32	–	–	–	–	–	–
County									
Bay	−0.45	−2.78	0.64	–	–	–	−0.41	−2.48	0.67
Broward	–	–	–	–	–	–	−0.38	−4.27	0.68
Citrus	–	–	–	0.62	2.99	1.87	–	–	–
Duval	−0.23	−3.15	0.79	−0.35	−4.68	0.70	−0.34	−3.78	0.71
Escambia	−0.30	−2.03	0.74	–	–	–	–	–	–
Lake	–	–	–	–	–	–	−0.36	−2.70	0.70

(continued)

Table 5. (continued)

Variable	Group A (under age 30)			Group B (ages 30–49)			Group C (age 50 or older)		
	Coeff.	*t*-stat	Odds ratio	Coeff.	*t*-stat	Odds ratio	Coeff.	*t*-stat	Odds ratio
Manatee	–	–	–	0.61	3.84	1.84	–	–	–
Marion	–	–	–	–	–	–	−0.58	−3.99	0.56
Miami-Dade	−0.26	−4.70	0.77	−0.45	−7.33	0.64	−0.57	−6.81	0.56
Monroe	–	–	–	0.29	2.67	1.34	0.24	2.26	1.28
Okaloosa	−0.44	−2.64	0.65	0.40	2.02	1.49	–	–	–
Palm Beach	0.18	2.22	1.20	–	–	–	–	–	–
Pasco	–	–	–	0.44	3.59	1.56	0.48	4.22	1.62
Pinellas	−0.32	−3.51	0.73	–	–	–	–	–	–
Site location									
Intersection	0.43	9.70	1.54	0.38	7.84	1.46	0.24	4.69	1.27
Driveway access	0.37	4.14	1.45	0.47	4.92	1.60	0.36	3.56	1.44
Parking lot	−0.47	−3.77	0.62	−0.76	−5.34	0.47	−0.46	−3.14	0.63
Motorcycle characteristics									
Body class									
Cruiser	−0.12	−2.35	0.89	0.14	2.74	1.15	0.26	5.89	1.30
Scooter	–	–	–	0.35	4.46	1.42	0.41	5.26	1.51
Sport	–	–	–	0.41	5.36	1.51	0.39	2.61	1.47
Supersport	0.18	4.51	1.20	0.35	5.88	1.42	–	–	–

equipment, assuming other variables are zero, puts his/her life at a high risk of incapacitating injury. Roadways with no shoulder increase the chance of more severe injury. Controlling traffic with a signal lowers the chance of severe injuries. Intersection or drive access increase the odds of serious injury in traffic crashes; this may be related to common right-of-way violations by other vehicles in which motorcyclists may not have enough time to react. The motorcycle injury severity level also changes both temporally and spatially. Florida DOT District 7 and some Florida counties (Pasco and Monroe) yield to a higher level of injury severity. Riding a motorcycle in early morning and/or on weekends puts young and middle-age motorcyclists (Groups A and B) at a higher risk of serious injury.

Stated Preference Survey

Whereas understanding the demographic characteristics of the population for traffic safety is essential, crash data provide very limited information about drivers and passengers. In this study, a survey was used to explore the characteristics of the motorcyclist segments identified by the motorcycle crash data.

A web-based survey with address-based sampling (ABS) was adopted to conduct a statewide survey. A total of 30,000 postcards were sent to randomly selected motorcyclists with a valid motorcycle endorsement who had registered at least one motor vehicle in Florida. In total, 976 survey responses, including partially completed surveys, were collected.

Table 6 summarizes the results of Chi-square tests and shows that there is a statistically significant association among the three groups and motorcycle use/riding/motorcycle types. Group A is more likely to ride sport/supersport-style motorcycles, and commuting is the primary purpose for riding compared to the other two groups. Of note is that about 17–25% of respondents indicated that they own a motorcycle but do not ride it regularly; this proportion is significantly higher in Group B compared to the other two groups.

The actual data of motorcycle use in the United States, such as VMT, is not readily available or the quality of existing information is questionable (*28*). Under current circumstances, it is difficult to collect adequate motorcycle exposure information by age group in the United States based on actual observation. This study used a series of questions to collect motorcycle exposure information of the three groups. Although the information is limited for estimating actual VMT for motorcycles in Florida, it provides insights to compare motorcycle exposure information, including riding frequency of the three groups. Table 7 shows the percentage within each cross-tab. For example, 18.3% of Group A reported riding a motorcycle more than 10,000 miles during the past 12 months, and riding almost every day. Cells with darker shading represent the higher percentages within the table. Motorcyclists in Group A rode more frequently than those in the other two groups, but almost the same proportion of Groups A and C reported that they rode 5,000 miles or more in the past 12 months. The majority of Group B rode more than once per week but not every

Table 6. Motorcycle Types and Primary Purpose of Riding by Age Group

	Group A (under age 30)	Group B (ages 30–49)	Group C (age 50 or older)	df	χ^2 (p-value)
Motorcycle riding					
I have a motorcycle and ride regularly	57.7%	59.1%	65.0%	8	17.737 (<.027)
I do not have a motorcycle but ride regularly as a passenger		0.6%	1.0%		
I have a motorcycle but do not ride a motorcycle regularly	17.5%	24.5%	19.5%		
I have a motorcycle endorsement but do not have a motorcycle and do not ride regularly	13.4%	11.6%	11.0%		
Total	100% (97)	100% (335)	100% (517)		
Primary purpose of riding					
Commuting	57.7%	25.2%	10.8%	8	120.838 (<.001)
Task-related trips		2.1%	3.8%		
Recreation	40.2%	67.8%	80.1%		
Competition		0.6%			
Other	2.1%	4.3%	5.4%		
Total	100% (97)	100% (326)	100% (502)		
Motorcycle type					
Cruiser	28.4%	41.4%	37.5%	14	229.202 (<.001)
On-/off-road	6.3%	1.5%	3.0%		
Scooter		3.9%	2.2%		
Sport/supersport	61.1%	26.1%	6.0%		
Standard	2.1%	3.3%	7.2%		
Touring/sport-touring		18.9%	37.1%		
Trike		0.9%	3.2%		
Other	2.1%	3.9%	3.8%		
Total	100% (95)	100% (333)	100% (501)		

Table 7. Riding Frequency and Estimated Annual Mileage

Group	Miles	Almost every day	More than 1× week	1× per week	A few times a month	A few times per year	Total
Group A (under age 30)	≤1,000 mile			4.9%	2.4%	20.7%	28.0%
	1,001–3,000 mile	2.4%	4.9%	4.9%	8.5%	2.4%	23.2%
	3,001–5,000 mile	2.4%	7.3%		2.4%		12.2%
	5,001–10,000 mile	11.0%	4.9%	2.4%			18.3%
	≥10,001 mile	18.3%					18.3%
	Total	34.1%	17.1%	12.2%	13.4%	23.2%	100.0%
Group B (ages 30–49)	Didn't ride in past 12 months			0.3%		5.3%	5.6%
	≤1,000 mile	0.3%	1.7%	4.3%	7.0%	14.3%	27.6%
	1,001–3,000 mile	0.3%	5.6%	3.3%	11.0%	2.0%	22.3%
	3,001–5,000 mile	2.7%	4.7%	4.3%	3.7%		15.3%
	5,001–10,000 mile	5.6%	7.6%	5.0%	1.0%		19.3%
	≥10,001 mile	6.3%	2.7%	0.3%	0.7%		10.0%
	Total	15.3%	22.3%	17.6%	23.3%	21.6%	100.0%
Group C (age 50 or older)	Didn't ride in past 12 months	0.2%	0.2%	0.2%		1.9%	2.6%
	≤1,000 mile or less	0.2%	0.9%	2.1%	7.5%	9.8%	20.6%
	1,001–3,000 mile	1.9%	6.1%	4.2%	10.5%	1.2%	23.8%
	3,001–5,000 mile	1.9%	6.3%	4.4%	4.0%	0.2%	16.8%
	5,001–10,000 mile	5.8%	12.9%	3.3%	3.0%	0.5%	25.5%
	≥10,001 mile	4.9%	4.9%	0.2%	0.5%	0.2%	10.7%
	Total	15.0%	31.3%	14.5%	25.5%	13.8%	100.0%

Table 8. Attitude Toward Training, Helmet Use, and Drinking

	Group A (under age 30)	Group B (ages 30–49)	Group C (age 50 or older)	df	χ^2 (*p*-value)
Did you complete formal motorcycle training to obtain your motorcycle endorsement?					
Yes	95.9%	82.6%	60.7%	4	77.318 (<.001)
No	4.1%	17.1%	38.5%		
I do not have a motorcycle endorsement		0.3%	0.8%		
Total	100% (97)	100% (327)	100% (499)		
In your opinion, how important is it for motorcyclists to be formally trained?					
Very important	71.1%	70.2%	69.6%	8	13.985 (0.082)
Fairly important	17.5%	12.2%	11.9%		
Important	4.1%	8.2%	11.9%		
Slightly important	7.2%	7.0%	4.3%		
Not important at all		2.4%	2.2%		
Total	100% (97)	100% (329)	100% (494)		
Which of the following best describes your use of a motorcycle helmet in Florida?					
I always wear a helmet when riding a motorcycle	80.4%	70.6%	65.7%	6	20.863 (0.002)
I wear a helmet most of the time	13.4%	15.5%	12.8%		
I wear a helmet on limited occasions (long-distance trips, winter season, and so forth)	6.2%	8.4%	11.2%		
I seldom or never wear a helmet		5.6%	10.4%		
Total	100% (97)	100% (323)	100% (492)		
In your opinion, how important is it for a motorcyclist to wear a helmet?					
Very important	80.2%	68.1%	60.9%	8	33.213 (<.001)
Fairly important	15.6%	9.6%	9.1%		
Important	2.1%	12.4%	14.4%		
Slightly important	2.1%	5.3%	9.3%		
Not important at all		4.6%	6.4%		
Total	100% (96)	100% (323)	100% (486)		
What kind of motorcycle helmet do you wear?					
Full-face	81.3%	54.6%	28.0%	10	119.001 (<.001)
Three-quarter	2.1%	14.4%	22.7%		
Modular		4.2%	7.5%		
Half	14.6%	22.4%	35.3%		
Off-road		0.6%	0.4%		
Other	2.1%	3.8%	6.0%		
Total	100% (97)	100% (323)	100% (492)		
In the past 30 days, how many times did you ride a motorcycle within 2 hours of drinking an alcoholic beverage?					
5 or more times		1.8%	1.6%	6	4.963 (.549)
3–4 times		2.4%	2.6%		
1–2 times	11.3%	8.5%	8.8%		
Never	88.7%	87.2%	87.0%		
Total	100% (97)	100% (328)	100% (491)		

day. It is also notable that more than 20% of Groups A and B reported that they rode just a few times per year.

Table 8 shows motorcyclist attitudes toward training, helmet use, and drinking by age group. Motorcyclists in all three groups strongly supported formal training, and approximately 96% of Group A stated that they obtained a motorcycle endorsement with formal motorcycle training. (Florida became a mandatory training state in July 2008; anyone seeking a new motorcycle endorsement in Florida is required to take a basic rider course.) The reported helmet usage and attitudes toward the importance of helmets are quite different among the three groups. It is notable that 95% of respondents in Group A stated that the use of a helmet is very important or fairly important, and 81% of respondents in this group reported wearing a full-face helmet. These measures (helmet usage and perceived safety value of helmet use) are lower in Group C compared to the other two groups. No difference was found in drinking and riding experience among the three groups.

As shown in Table 9, each age group displays significant differences in motorcycle club membership and bike event attendance. More than half of motorcyclists in Groups A and B have not attended a major bike event during the past 12 months, but 30% of Group C stated that they attended at least two or more bike events. Of

Table 9. Motorcycle Club Membership, Bike Events, and Preferred Means of Contact

	Group A (under age 30)	Group B (ages 30–49)	Group C (age 50 or older)	df	χ^2 (p-value)
Are you a member of a riding group?					
Yes	11.1%	11.3%	28.0%	2	38.810 (<.001)
No	88.9%	88.8%	72.0%		
Total	100% (90)	100% (320)	100% (485)		
Did you attend any major bike events in Florida during the past 12 months?					
Never	52.0%	54.0%	43.5%	4	28.002 (.002)
Once	25.5%	23.4%	27.2%		
Twice	14.7%	13.6%	12.1%		
More than twice	7.8%	8.9%	17.1%		
Total	100% (90)	100% (320)	100% (485)		
What would be the best way to reach you about motorcycle safety information in Florida?					
Direct mail	34.0%	23.2%	19.5%	20	80.114 (<.001)
Billboards	6.4%	7.7%	5.6%		
Electronic overhead highway sign	18.1%	12.1%	7.7%		
Television	6.4%	12.7%	20.1%		
Radio	7.4%	6.2%	1.7%		
Telephone		0.6%	1.0%		
Social network websites	11.7%	15.8%	8.5%		
E-mail	13.8%	15.5%	25.5%		
Online rider forums		1.9%	2.9%		
Newspaper			1.7%		
Other	2.1%	4.3%	6.0%		
Total	100% (94)	100% (323)	100% (483)		
How often do you log into social media networks (e.g., Facebook, Google + , and so forth)?					
More than once per day	50.0%	44.4%	31.3%	8	29.641 (<.001)
About once per day	26.7%	20.9%	21.6%		
A few times per week	10.0%	11.6%	19.4%		
A few times per month		5.0%	6.2%		
Less than a few times per month	13.3%	18.2%	21.6%		
Total	100% (90)	100% (320)	100% (485)		

note is that all three groups expressed significant interest in getting information by direct mail. Although more than 50% of respondents in all three groups indicated that they log into social media at least once per day, they indicated that social media was not the best way to receive information about motorcycle safety.

The results of the survey demonstrated that the three groups are significantly different in many categories, such as riding purpose, exposure, attitudes toward risk factors, and common activities. Although the survey adopted random sampling, the interpretation of findings is limited by the low response rate and the particularly small sample size of Group A.

Discussion

Considering that motorcycles have significantly higher fatality rates per registered vehicle or per estimated VMT compared to passenger cars, they are the most dangerous means of transportation in the United States, and the sharp increase in motorcycle registrations has resulted in a significant increase in motorcycle fatalities as well as major changes in the demographics of motorcyclists. Florida has about 10% of all motorcycle registrations in the United States and saw similar demographic changes. In this study, Florida motorcycle crash data were divided into three clusters based on the age of motorcycle operators. The age range for each cluster was determined based on age thresholds of the increase/decrease in serious and fatal motorcycle crashes during the past 10 years. Descriptive analyses including Chi-square tests were used to evaluate crash characteristics of the three groups. Each group has diverse crash characteristics as well as different levels of exposure to common risk factors, including no helmet use, suspected alcohol use, and speeding.

Crash injury models were developed to understand variable increases in the level of injury severity in each group. Risk factors such as no helmet use and suspected alcohol use appeared in all three models as a significant contributing factor. In addition, curves with level terrain, curves with rolling terrain, intersections, and drive access are common variables that increase the level of injury severity in all three age groups. Speeding (riding a motorcycle 20 mph over the posted speed limit) was much more prevalent in Group A, but motorcyclists in Groups B and C were more likely to suffer increased injury severity by riding a motorcycle on a roadway with a higher posted speed limit (55 mph or higher). Similarly, riding a motorcycle in dark conditions without street lighting increased injury severity for Groups B and C only. Overall, the crash injury models confirmed that many variables that increased the level of injury severity in motorcycle crashes were applicable to all three groups, whereas the frequency of some behaviors varied by group.

A survey was conducted to collect information not available in crash data, such as attitudes and behavioral characteristics of each segment. The three groups are significantly different in riding purpose, primary motorcycle type, riding frequency, and annual miles. Although there is a significant difference in formal training experience among groups, all three demonstrated positive attitudes toward formal motorcycle training. All three groups were very different concerning type of helmet, reported helmet use, and attitudes toward wearing a helmet. The three groups were also different in motorcycle club membership, bike event attendance, and preferred means of communication.

Overall, the three motorcycle segments identified in this study have distinctive riding characteristics, crash outcomes, and attitudes toward risk factors. This is essential information for developing and managing motorcycle safety programs in an effective and efficient manner. Current and past motorcycle safety interventions have been limited to general public campaigns, which target the whole population of either motorcyclists or motorists. Consideration of a tailored approach based on target audience segmentation warrants further investigation, given its widespread success in other areas of public health. Targeted segmentation has been used in public health settings to develop effective messages and intervention strategies and to choose the most appropriate communication channels for each segment. Target audience segmentation requires planners to divide the general target population into homogeneous subgroups, as each resulting subgroup or segment has unique characteristics compared to other segments, and shared beliefs, attitudes, and behaviors about a topic. Individuals within a segment are likely to show a similar response to an intervention approach tailored for that segment. Grier and Kumanyika (*29*) call for a targeted marketing approach for the promotion of public health products with proven benefits, such as motorcycle helmets. The field of marketing has moved beyond universal campaigns to develop techniques to target specific groups of consumers with shared attributes. When marketing messages are customized for subgroups of the population that have been segmented according to relevant demographic and behavioral characteristics, these messages are more favorably received and more likely to elicit the intended response than mass-marketed messages.

Conclusion

The primary focus of many previous motorcycle safety studies has been investigating and identifying contributing factors to motorcycle crashes and injury outcomes. These study findings are useful but can be limited for developing effective behavioral traffic safety programs, as they do not include a complete understanding of the primary target audience. A possible explanation is that most researchers work with crash data that often provide detailed information about motorcycle crashes but only limited information about vehicle operators.

This study showed that motorcycle fatalities have increased in both Group A (under age 30) and Group C (age 50 or older), both of which have undergone significant demographic changes during the past 10 years. A statewide survey was used to explore the characteristics of motorcyclists in each segment; the stated preference survey was an effective tool to investigate the attitudes and behavioral information of motorcyclists in each segment. The findings were compared and incorporated with the analysis of crash data. All three motorcyclist segments have different riding styles and behavioral characteristics, which can affect crash outcomes at different levels. This study examined motorcyclist age as an important evaluation variable for classifying motorcyclists, and demonstrated that the selected age thresholds establish three significant motorcyclist segments.

Further research is needed to understand motorcycle segments beyond age, direct relationships between segments to crash outcomes, and effective intervention programs reflecting each segment's unique characteristics.

Author Contributions

The authors confirm contribution to the paper as follows: study conception and design: CL; data collection: CL, BK; analysis and interpretation of results: CL, BK, SJ; draft manuscript preparation: CL, BK, SJ, VS. All authors reviewed the results and approved the final version of the manuscript.

References

1. National Center for Statistics and Analysis. *Motorcycle 2015 Data (Updated)*. Traffic Safety Facts, Report No. DOT HS 812 353.National Highway Traffic Safety Administration, Washington, D.C., 2017.
2. Retting, R., and H. Rothenberg. *Motorcyclist Traffic Fatalities by State: 2015 Preliminary Data*. Governors Highway Safety Association, Washington, D.C., 2016.

3. Xu, G., R. Crowe, G. Rousseau, and L. K. Bedsole. *Slamming on the Brakes on a Mounting Problem*. FHWA-HRT-17-004. US Department of Transportation, Washington, D.C., 2002.
4. McCartt, A. T., L. Blanar, E. R. Teoh, and L. M. Strouse. Overview of Motorcycling in the United States: A National Telephone Survey. *Journal of Safety Research*, Vol. 42, No. 3, 2011, pp. 177–184.
5. Shankar, U., and C. Varghese. *Recent Trends in Fatal Motorcycle Crashes: An Update (DOT HS 810 606)*.National Highway Traffic Safety Administration, Washington, D.C., 2006.
6. Mattsson, M., and H. Summala. With Power Comes Responsibility: Motorcycle Engine Power and Power-to-Weight Ratio in Relation to Accident Risk. *Traffic Injury Prevention*, Vol. 11, 2010, pp. 87–95.
7. Rolison, J. J., P. J. Hewson, E. Hellier, and L. Hurst. Risks of High-Powered Motorcycles Among Younger Adults. *American Journal of Public Health*, Vol. 103, 2013, pp. 568–571.
8. Teoh, E. R., and M. Campbell. Role of Motorcycle Type in Fatal Motorcycle Crashes. *Journal of Safety Research*, Vol. 41, No. 6, 2010, pp. 507–512.
9. Bjornskau, T., R. O. Naevestad, and J. Akhtar. Traffic Safety Among Motorcyclists in Norway: A Study of Subgroups and Risk Factors. *Accident Analysis & Prevention*, Vol. 49, 2012, pp. 50–57.
10. Harrison, W. A., and R. Christie. Exposure Survey of Motorcyclists in New South Wales. *Accident Analysis & Prevention*, Vol. 37, 2005, pp. 441–451.
11. Dischinger, P. C., G. E. Ryb, S. M. Ho, and C. A. Burch. The Association Between Age, Injury and Survival to Hospital among a Cohort of Injured Motorcyclists. *Annual Proceedings, Association for the Advancement of Automotive Medicine*, Vol. 51, 2007, pp. 97–110.
12. Lee, C., J. Pino, and K. Choi. Lessons Learned from Motorcyclist Surveys: Riders' Attitudes and Behaviors in Florida. *Transportation Research Record: Journal of the Transportation Research Board*, 2013. 2388: 71–78.
13. Stutts, J., R. Foss, and C. Svoboda. Characteristics of Older Motorcyclist Crashes. *Annual Proceedings, Association for the Advancement of Automotive Medicine*, Vol. 48, 2004, pp. 197–211.
14. Brown, J. B., P. E. Bankey, J. T. Gorczyca, J. D. Cheng, N. A. Stassen, and M. L. Gestring. The Aging Road Warrior: National Trend Toward Older Riders Impacts Outcome After Motorcycle Injury. *American Surgeon*, Vol. 76, No. 3, 2010, pp. 279–286.
15. Centers for Disease Control and Prevention. Trends in Motorcycle Fatalities Associated with Alcohol-Impaired Driving: United States, 1983–2003. *Morbidity and Mortality Weekly Report*, Vol. 53, No. 47, 2004, pp. 1103–1106.
16. Jung, S., Q. Xiao, and Y. Yoon. Evaluation of Motorcycle Safety Strategies Using the Severity of Injuries. *Accident Analysis & Prevention*, Vol. 59, 2013, pp. 357–364.
17. Maistros, A., W. H. Schneider, and P. T. Savolainen. A Comparison of Contributing Factors Between Alcohol-Related Single Vehicle Motorcycle and Car Crashes. *Journal of Safety Research*, Vol. 49, 2014, pp. 129–135.
18. Schneider, W. H., P. T. Savolainen, D. V. Boxel, and R. Beverley. Examination of Factors Determining Fault in Two-Vehicle Motorcycle Crashes. *Accident Analysis & Prevention*, Vol. 45, 2012, pp. 669–676.
19. Eustace, D., and V. K. Indupuru. A Simplified Method for Analyzing Factors Contributing to Motorcyclists' Fatal Injuries in Ohio. *Advances in Transportation Studies*, Vol. 24, 2011, pp. 85–96.
20. Nunn, S. Death by Motorcycle: Background, Behavioral, and Situational Correlates of Fatal Motorcycle Collisions. *Journal of Forensic Sciences*, Vol. 56, No. 2, 2011, pp. 429–437.
21. Schneider, W. H., and P. T. Savolainen. Comparison of Severity of Motorcyclist Injury by Crash Types. *Transportation Research Record: Journal of the Transportation Research Board*, 2011. 2265: 70–80.
22. Savolainen, P., and F. Mannering. Probabilistic Models of Motorcyclists' Injury Severities in Single- and Multi-Vehicle Crashes. *Accident Analysis and Prevention*, Vol. 39, No. 5, 2007, pp. 955–963.
23. Sauter, C., S. Zhu, S. Allen, S. Hargarten, and P. M. Layde. Increased Risk of Death or Disability in Unhelmeted Wisconsin Motorcyclists. *Wisconsin Medical Journal*, Vol. 104, No. 2, 2005, pp. 39–44.
24. Shankar, V., and F. Mannering. An Exploratory Multinomial Logic Analysis of Single-Vehicle Motorcycle Accident Severity. *Journal of Safety Research*, Vol. 27, No. 3, 1996, pp. 183–194.
25. Talving, P., P. G. R. Teixeira, G. Barmparas, J. DuBose, C. Preston, K. Inaba, and D. Demetriades. Motorcycle-Related Injuries: Effect of Age on Type and Severity of Injuries and Mortality. *Journal of Trauma-Injury Infection and Critical Care*, Vol. 68, No. 2, 2010, pp. 441–446.
26. Mulvihill, C. M., and N. L. Haworth. *A Survey of the Crash Characteristics of Older Motorcyclists*. Monash, Australia: Monash University Accident Research Centre, 2005.
27. Quddus, M. A., R. B. Noland, and H. C. Chin. An Analysis of Motorcycle Injury and Vehicle Damage Using Ordered Probit Models. *Journal of Safety Research*, Vol. 33, No. 4, 2002, pp. 445–462.
28. Middleton, D., P. Turner, H. Charara, S. Sunkari, S. Geedipally, and R. Scopatz. *NCHRP Report 760: Improving the Quality of Motorcycle Travel Data Collection*. Transportation Research Board, Washington, D.C., 2016.
29. Grier, S. A., and S. Kumanyika. Targeted Marketing and Public Health. *Annual Review of Public Health*, Vol. 31, 2010, pp. 349–369.

The Standing Committee on Motorcycles and Mopeds (ANF30) peer-reviewed this paper (18-05572).

Article

TRR
JOURNAL OF THE TRANSPORTATION RESEARCH BOARD

Transportation Research Record
2018, Vol. 2672(34) 122–134

DOI: 10.1177/0361198118797212
journals.sagepub.com/home/trr
SAGE

Using Deep Learning in Severity Analysis of At-Fault Motorcycle Rider Crashes

Subasish Das[1], Anandi Dutta[2], Karen Dixon[1], Lisa Minjares-Kyle[3], and George Gillette[4]

Abstract

Motorcyclists are vulnerable highway users. Unlike passenger vehicle occupants, motorcycle riders do not have either protective structural surrounding or the advanced restraints that are mandatory safety features in cars and light trucks. Per vehicle mile traveled, motorcyclist fatalities occurred 27 times more frequently than passenger car occupant fatalities in traffic crashes. In addition, there were 4,976 motorcycle crash-related fatalities in the U.S. in 2014—more than twice the number of motorcycle rider fatalities that occurred in 1997. It shows that, in addition to current efforts, research needs to be conducted with additional resources and in newer directions. This paper investigated five years (2010–2014) of Louisiana at-fault motorcycle rider-involved crashes by using deep learning, which is a competent tool for mapping a high-multidimensional input into a smaller multidimensional output. The current study contributes to the existing injury severity modeling literature by developing a deep learning framework, named as *DeepScooter*, to predict motorcycle-involved crash severities. The final deep learning model can predict severity types with 100% accuracy with training data, and with 94% accuracy with test data, which is not attainable by using a statistical method or machine learning algorithm. The intensity of severities was found to be more likely associated with rider ejection, two-way roadways with no physical separation, curved aligned roadways, and weekends. It is anticipated that the *DeepScooter* framework and the findings will provide significant contributions to the area of motorcycle safety.

Typically identified as "vulnerable roadway users," motorcyclists are an at-risk group of roadway users whose death rates have consistently remained higher than other vehicle-based roadway users. In 2014 alone, motorcyclist fatalities "occurred 27 times more frequently than fatalities in other vehicles" with 4,976 motorcycle crash-related fatalities in the U.S. This is more than twice the number of motorcycle rider fatalities that occurred in 1997, and contrasts the 27% reduction in the number of fatalities involving passenger cars and light trucks (*1*).

Motorcycle crashes represent 5% of all fatal crashes in Louisiana each year, and yet only represent 1% of all total crashes. In 2016, 98 fatal crashes (30% higher than 2010 statistics) involved motorcycles (*2*). The conventional approach to crash-severity analysis has been to establish associations between traffic and driver characteristics, roadway and environment conditions, and crash occurrence. The shortcoming of most of the models developed using this approach is that they rely on aggregate measures and general inferences. In addition, identifying contributing factors using observational data comprises a wide variety of associations because of the assumptions considered during different modeling techniques. Thus, it is essential to determine the influence of patterns of associated factors with crash severity. The 2010–2014 traffic crash data associated with at-fault motorcycle riders for Louisiana were used in this study. This study restricted the deep learning framework development by analyzing only at-fault motorcycle rider crashes because not-at-fault motorcycle rider crashes involve other vehicle drivers' driving traits, which is outside the scope of the current study. A deep learning framework "*DeepScooter*" was developed to better predict injury outcomes from the significant factors associated with motorcycle crashes.

[1]Texas A&M Transportation Institute, Texas A&M University System, College Station, TX
[2]Computer Science and Engineering Department, Texas A&M University, College Station, TX
[3]Texas A&M Transportation Institute, Texas A&M University System, Houston, TX
[4]Department of Civil and Environmental Engineering, University of California, Berkeley, CA

Corresponding Author:
Address correspondence to Subasish Das: s-das@tti.tamu.edu

Earlier Work and Research Context

The most prominent area of transportation safety analysis are crash frequency analysis (count data problem), and injury-severity analysis (classification problem). Lord and Mannering provided a detailed research synthesis on crash count data and related methods and limitations for examining such data (*3*). Savolainen et al. presented a similar assessment on injury-severity analysis (*4*). Recently, Mannering and Bhat extended and bridged both of these studies (*5*). Mannering et al.'s study presented a detailed discussion of unobserved heterogeneity in crash data analysis along with their strengths and weaknesses (*6*).

Research and analysis of motorcycle crashes throughout the years has provided an abundance of valuable and important information to assist with the overall need to reduce motorcyclists' injuries and deaths caused by crashes. Rider variables such as age, gender, and impairment, for example, have consistently been identified in the literature to influence the likelihood of increased crash severity. Researchers have identified significant relationships between the age of a rider and increased crash-severity outcomes in which older riders (over the age of 25), although less likely to be at fault in a collision, are more likely to be severely injured in crashes compared with younger riders (*7–13*). Male riders have lower probabilities of being severely injured in a crash compared with females, but on average have a higher likelihood of being fatally injured in a crash (*7*, *8*, *14*).

Intoxicated driving has been identified as a significant contributor to motorcycle crash severity. Crashes involving impaired riders have been found to be more severe with a higher risk of fatal or major injury outcomes than crashes involving non-impaired riders (*10*, *15–18*). Research has shown that impaired motorcyclists are more likely to be involved in single-vehicle crashes, be at fault, and less likely to be wearing a helmet at the time of the crash, which can affect crash severity (*9–10*, *13*, *15*).

Helmet use has been identified throughout several studies as a significant variable in motorcycle crash severity, in which the likelihood of a fatality or the severity of a crash is higher when a helmet is not worn (*8–10*, *13*, *14*). Crashes in which a motorcyclist is wearing a helmet have been associated with reduced chances of severe or fatal injuries in both rural or urban environments, decreases in injuries sustained in both single and multi-vehicle crashes, and reduced frequency of severe injuries (*7*, *12–14*, *16*, *19*).

Temporal factors such as season, day of the week (weekend vs. weekday), and time of day at which motorcycle crashes occurred have also been linked to motorcycle crash-severity outcomes in the literature. Motorcycle crashes that occur at night (between the hours of 8 p.m. and 6 p.m.) or early morning hours, on weekends, during the early months of the rider "season" or summer months have been shown to have higher fatal and severe injury probabilities (*10*, *12–14*, *20*, *21*). For example, Savolainen and Mannering found that crashes that occurred in April and July had a 111% and 98% greater probability of being fatal. Later months also showed a lower likelihood of protective gear use because of higher temperatures (*14*).

Several geometric variables have been identified in the literature to contribute to motorcycle crash-severity outcomes including roadway type, light conditions, posted speeds, roadway features such as curves or T-junctions, and weather. According to the literature, roadways that have the following characteristics significantly influence motorcycle crash severity: two-lane roads, farm to market roads, curved roads, looped or "lollipop" designed roads, and T-junction roads controlled by "stop, give-way signs or markings," (*7*, *9*, *10*, *16*, *19–22*). Motorcycle crashes that occurred on highways, near driveways, intersections, or signalized intersections were also found to influence crash severity, as it is estimated that these factors negatively affect a driver's ability to see a motorcyclist (*9*, *23*). Crashes that occurred in rural localities were also significantly linked to motorcycle crash severity, which may be because of overall higher posted speeds, roadway geometry, the absence of streetlights, and a higher propensity for two-lane roadways (*16*, *18*).

Motorcycle injury severity has also been significantly linked to speed. Roadways with higher posted speeds or motorcycle crashes that occur at "unsafe speeds," where speed is listed as the primary contributing factor or posted speed limits are higher than 55 mph, have been shown to increase the probability of a fatality and more severe injuries (*10*, *18*, *20*, *22*). For example, Savolainen and Mannering found a 212% increase in the probability of a fatality when the crash citation involved the factor "unsafe speed" (*6*). This is true for both single-vehicle crashes and multi-vehicle crashes (*13*, *14*, *18*).

Motorcycle crashes have been found to be more severe during daytime weather versus wet or rainy weather (*7*, *10*, *13*). Savolainen and Mannering posit this may be as a result of lower speeds maintained by riders while driving through wet or rainy conditions, in which riders may exhibit more caution and lower speeds as they adjust to "perceived higher risk" on wet roadways versus dry (*7*).

Overview of Models Used to Analyze Crash Severity

Several modeling approaches have been applied throughout the literature to analyze crash severity by examining injury-severity levels related to the various factors discussed in the previous paragraphs. This analysis will provide a snapshot of models used in the literature review and background of the research works in this paper.

Many studies have used multinomial ordered probit and logit models for crash-severity analysis, which analyzes various categories of injury-severity levels in order of severity ranging from no injury to a fatality (*11, 12, 17, 21–23*). Research has identified potential limitations to this model because of constrained effects resulting from ordered modeling and underreporting of crashes that have minor or no severe injuries (*7*).

Multinomial logit models (MNL) have been used by researchers to address the above issues as MNL models "consider three or more outcomes and do not explicitly consider the ordering," (*23*). Researchers have used this model to analyze motorcycle rider accident severity in single-vehicle crashes (*15*), develop probabilistic models of motorcyclists' injury severities for single and multi-vehicle crashes (*7*), compare severity of motorcycle injury by crash types (*19*), analyze differences in factors that affect severity of motorcyclist's crash injuries (*16*), and predict the probability of crash severities (*10*).

Mixed logit analysis models allow for "heterogeneous effects and correlation in unobserved factors" (*19*) and have been used to examine crash-specific factors (crash factors, roadway and environment conditions, and vehicle attributes) on two-vehicle crash-severity outcomes involving motorcycles (*14*).

Log linear modeling provides "measures of the magnitude, direction and statistical significations of main effects" and "interactions among a set of categorical variables." Haque et al. utilized log linear modeling to investigate effects of traffic, environmental, and roadway factors on roadway crash in Singapore to establish influential factors in various location types (*24*). Other methodological approaches include Empirical Bayesian analysis and stepwise logic regression (*8, 25, 26*). It is important to note that conventional statistical models are good at statistical inference. The common limitation of these methods is poor prediction accuracy. In addition, the models are based on assumptions. Violations of any of the assumption will produce biased results.

Overview of Deep Learning Method Analysis

Within the field of data analysis, there are two significantly differing opinions regarding its appropriate treatment—roughly categorized as data or statistical modeling and algorithmic modeling or machine and deep learning (*27*). The data modeling approach transcribes to the belief that an underlying stochastic process has generated the data, such that the response variables can be related to a set of predictor variables. In order to evaluate the model's success, this approach would apply goodness-of-fit tests based around established and acceptable margins of certainty. In contradiction, algorithmic modeling focuses on the process of understanding the unknown by minimizing the error rate through a black box algorithmic model.

Although deep learning has not been widely used in past studies, many studies have used machine learning and data mining algorithms in transportation safety analysis (*28–40*). Deep learning is a branch of artificial intelligence that attempts to model complex information through a series of processing layers. Although deep learning has recently become the new primary focus within the artificial intelligence community, the models and ideas behind this technique have been discussed for over half a century.

In the field of transportation engineering, the majority of statistical work could be shown to fall into the category of data modeling. Although there are not similar papers in deep learning among the established literature for motorcycle safety, there are examples of how deep learning can be applied to other transportation engineering problems through traffic data imputation, short-term traffic flow prediction, vehicle classification, and sustainable guideline development. Duan et al. highlight the importance of clean and complete traffic datasets, as the growing sensor infrastructure is generating ubiquitous data for analysis (*41*). While Duan et al. were interested in the imputation of missing data, Polson et al. focused on the prediction given short-term conditions (*42*). Yu et al. developed a fine-grained vehicle classification approach that applied a convolutional neural network with a joint Bayesian network to classify a vehicle similar to the methods applied for classifying a face (*43*).

Data Processing

The master database created for this analysis includes 10,099 motorcycle-involved crashes from the police-reported crash data in 2010–2014, Louisiana. This study mainly focused on at-fault motorcycle riders. Louisiana crash data contains a variable named "*Vehicle Number*," in which 1 denotes at-fault riders. The final dataset of at-fault motorcycle crashes included 6,853 crashes. This study analyzes only at-fault motorcycle rider-involved crashes because not-at-fault rider crashes involve other vehicle drivers' driving traits, which is outside the scope of the current study. The severity of crashes is recorded as five injury levels (commonly known as KABCO injury scale): fatality (K), incapacitating injury (A), non-incapacitating injury (B), possible/complaint injury (C), and no injury (O). The fatal injury category includes crashes that result in death within 30 days of the crash. The incapacitating injury prevents the injured person from normal daily work. The non-incapacitating injury includes evidence of significant injury during police reporting. The possible injury indicates complaints of pains or stresses with no physical evidence. No-injury crashes, also known as Property Damage Only (PDO), do not involve any injury.

To determine important variables, past studies have been consulted. A wider list of variables has been selected in the preliminary analysis. As a result of lack of adequate information (for example, blood alcohol content level of motorcycle riders), some of the variables were not explored for the final analysis. The final dataset contains 16 predictors of the dataset with 79 levels, of which three are numerical, and 13 are categorical (76 levels). The general findings from the numerical variables are following:

- The median value of crash hour is 4 p.m., which may be because of the peak-hour traffic volume increase at the end of the workday.
- The median number of vehicles was two, indicating that motorcycle crashes more often are not single-vehicle crashes.
- Lastly, the median rider age was 39, which may be because of the higher number of middle-aged motorcycle riders compared with younger riders. Regardless, young riders have been found to have a higher crash likelihood compared with older riders.

Table 1 lists descriptive statistics of the categorical variables used in the final analysis. As this study focuses on at-fault motorcycle rider crashes, the number of crashes and the number of at-fault motorcycle riders are same. Of the total 6853 crashes, 4.6% crashes were classified as fatal crashes, 74.1% are classified as injury crashes and the rest of these crashes are PDO. Sixty-two percent of crashes occurred more frequently on two-lane rural roadways that did not have a dividing barrier, which is similar to previous findings in the literature. Motorcycle crashes were also more likely to occur on straight level roadways (74.9%), but it is important to note that 15.6% occurred on curve level roadways, a feature which has been shown to impact on crash severity. A majority of crashes occurred during daytime and clear weather conditions, which correlates with previous findings that suggest higher speeds are more common during clear weather. Motorcycle crashes were more likely to occur over the weekend (Friday–Sunday) and over half of crashes analyzed, 52.3%, involved a rider being thrown from the motorcycle, which is likely because a motorcycle offers a rider no protection. Localities with business entities show high likelihood of motorcycle crash involvements. Lastly, nearly 27% of crashes involved "driver inattention" compared with only 3.6% of crashes that involved alcohol-impaired driving.

Deep Learning

History

The history of artificial neural networks dates back to 1950. Rosenblatt's Perceptron algorithm was the earliest example of an artificial neural network (ANN). In the

Table 1. Descriptive Statistics of Categorical Variables

Category	Count	%
Road condition		
No abnormalities	6374	93.0%
Animal in roadway	110	1.6%
Construction, repair	75	1.1%
Loose surface material	68	1.0%
Object in roadway	48	0.7%
Other types	178	2.6%
Roadway type		
Two-way road with no physical separation	4247	62.0%
Two-way road with a physical separation	1667	24.3%
One-way road	687	10.0%
Two-way road with a physical barrier	218	3.2%
Other types	24	0.5%
Highway type		
State hwy	2751	40.1%
City street	1504	21.9%
U.S. hwy	1092	15.9%
Parish road	887	12.9%
Interstate	615	9.0%
Toll road	4	0.1%
Locality type		
Business continuous	1931	28.2%
Business, mixed residential	1854	27.1%
Residential district	958	14.0%
Residential scattered	950	13.9%
Open country	844	12.3%
Manufacturing or industrial	145	2.1%
Other types	171	2.5%
Alignment type		
Straight-level	5131	74.9%
Curve-level	1066	15.6%
Straight-level-elevated	142	2.1%
On grade-curve	137	2.0%
Curve-level-elevated	132	1.9%
On grade-straight	122	1.8%
Other types	123	1.8%
Lighting condition		
Daylight	4733	69.1%
Dark—continuous street light	1025	15.0%
Dark—no street lights	687	10.0%
Dark—street light at intersection only	194	2.8%
Dusk	151	2.2%
Dawn	51	0.7%
Other types	12	0.2%
Weather condition		
Clear	5762	84.1%
Cloudy	820	12.0%
Rain	206	3.0%
Fog/smoke	34	0.5%
Other types	31	0.4%
Collision type		
Single vehicle	2578	37.6%
Rear end	1317	19.2%
Right angle	908	13.2%
Left turn—opposite direction	445	6.5%
Sideswipe—same direction	442	6.4%
Other types	1163	17.0%
Day of the week		
Saturday	1352	19.7%
Sunday	1114	16.3%

(continued)

Table 1. (continued)

Category	Count	%
Friday	1099	16.0%
Thursday	980	14.3%
Wednesday	801	11.7%
Tuesday	768	11.2%
Monday	739	10.8%
First harmful event		
Motor vehicle in transport	4260	62.2%
Ran off road right	680	9.9%
Overturn/rollover	405	5.9%
Other non-collision	309	4.5%
Crossed median/centerline	241	3.5%
Ran off road left	239	3.5%
Other types	719	10.5%
Rider ejection		
Totally ejected	3584	52.3%
Not ejected	2665	38.9%
Partially ejected	451	6.6%
Unknown	153	2.2%
Rider condition		
Normal	3968	57.9%
Inattentive	1785	26.0%
Unknown	588	8.6%
Drinking alcohol—impaired	244	3.6%
Distracted	119	1.7%
(Other)	149	2.2%
Rider severity		
Possible/complaint	2319	33.8%
Non-incapacitating/moderate	2280	33.3%
No injury	1460	21.3%
Incapacitating/severe	478	7.0%
Fatal	316	4.6%

earlier stages, Perceptron failed to approximate many nonlinear decision functions. A solution was developed later by stacking multiple layers of linear classifiers, known as multilayer perceptron, to approximate nonlinear decision functions. Because of a lack of computational power, ANN faced a slower pace of development during 1990–2000. Since late 2000, ANN has seen explosive progress as a result of parallel processing. ANN processes many parameters and approximates nonlinear functions. For a very deep and complex problem, ANN can provide a reasonable estimate. This new branch of complex problem solving is known as "deep learning" Many researchers consider learning to be deep if ANN has more than two layers. There are other interpretations of the word "deep," including: (1) deep learning can provide solutions for unlabeled data, and (2) deep means autonomous (*44–47*). Figure 1*a* shows the timeline of the scientific evolution of deep learning. This study used the 'R with H2O.ai' platform to perform the analysis (*48*).

Theory

Interested readers can consult these references for a better understanding of deep learning architecture (*43–46*). For the sake of easy interpretation, only a brief overview of deep learning concept is provided here. Consider a set of data points $\{x^{(1)}, x^{(2)}, \ldots, x^{(m)}\}$ in which each data point has many dimensions. These data can be mapped to another set of data points $\{z^{(1)}, z^{(2)}, \ldots, z^{(m)}\}$, where zs have lower dimensionality than xs. In place of using high-dimensional x, low-dimensional z can reconstruct x. To map data back and forth, a relationship can be developed:

$$z^{(i)} = W_1 x^{(i)} + b_1 \tag{1}$$

$$\tilde{x}^{(i)} = W_2 x^{(i)} + b_2 \tag{2}$$

If x^i is a two-dimensional vector, it is possible to visualize the data to find W_1, b_1 and W_2, b_2 analytically as the experiment above suggested. For high-dimensional data, the visualization is not possible. As the target is to attain $\tilde{x}^{(i)}$ to estimate $x^{(i)}$, an objective function can be used:

$$\begin{aligned} J(W_1, b_1, W_2, b_2) &= \sum_{i=1}^{m} \left(\tilde{x}^{(i)} - x^{(i)}\right)^2 \\ &= \sum_{i=1}^{m} \left(W_2 x^{(i)} + b_2 - x^{(i)}\right)^2 \\ &= \sum_{i=1}^{m} \left(W_2\left(W_1 x^{(i)} + b_1\right) + b_2 - x^{(i)}\right)^2 \end{aligned} \tag{3}$$

which can be minimized using stochastic gradient descent. This concept is also known as a linear autoencoder (as shown in Figure 1*b*).

An example of a deep network with two hidden layers (W_1, *and* W_2) is shown in Figure 1*b*. To train the neurons, an autoencoder (with parameters W_1 and W_1') can be trained. Later W_1 will be used to compute the values for the neurons for all data, which will then be used as input data to the subsequent autoencoder. This autoencoder uses the values for the neurons as inputs, and trains an autoencoder to predict those values by adding a decoding layer with parameters W_2'.

Model Development

The current study has developed a deep learning framework named "*DeepScooter*." This framework is based on five major steps: (1) finalize the dataset by selected contributing factors, (2) divide the dataset into training, validation, and text data, (3) use a basic deep learning model and check model accuracies for training, validation, and text data, (4) perform tuning to get better estimates with higher accuracies, and (5) select the final model. Figure 2 illustrates the general framework of *DeepScooter*.

A detailed step-by-step procedure is described below:

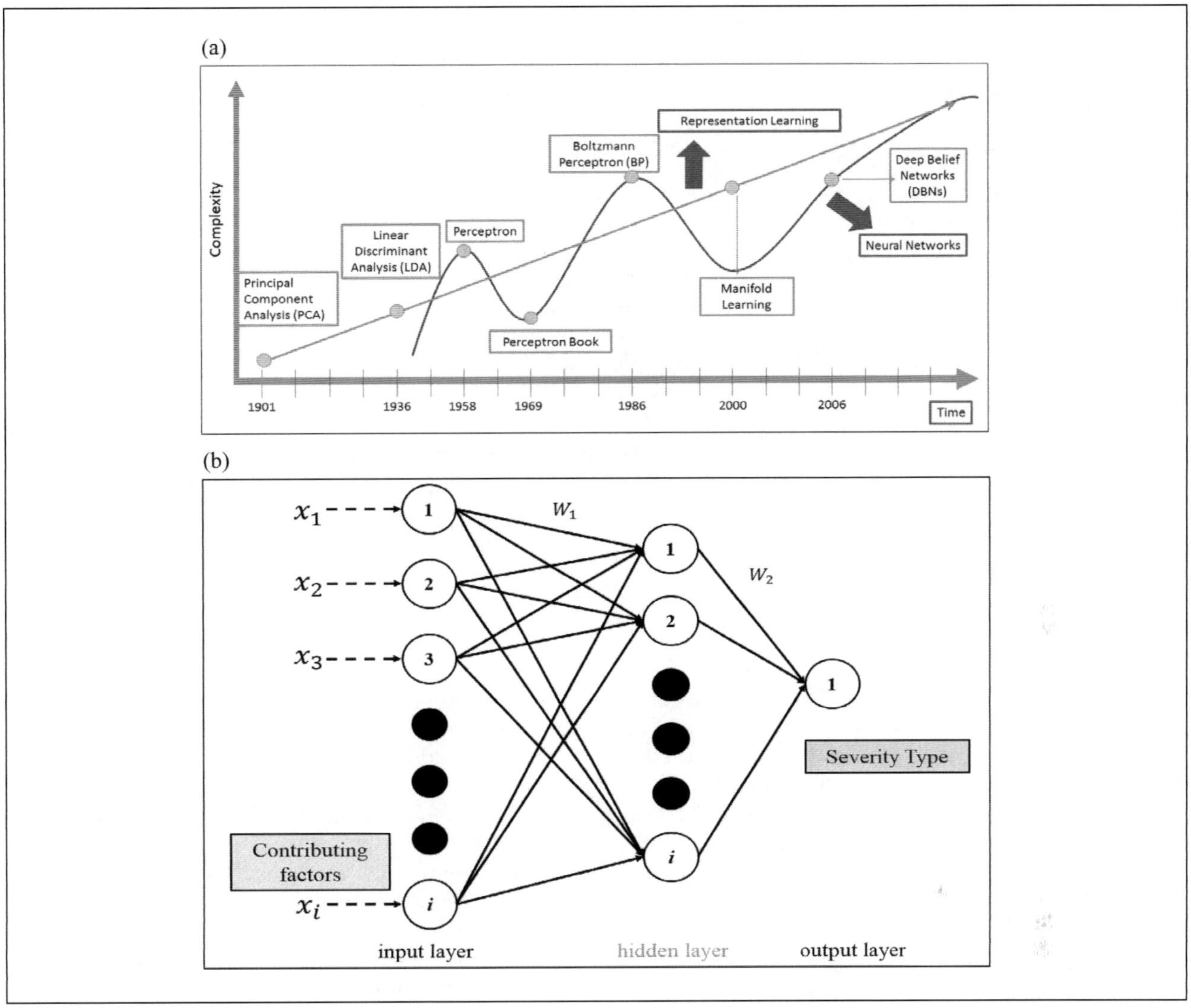

Figure 1. (*a*) Timeline of deep learning (reproduced from *44*) and (*b*) deep learning algorithmic concept.

- The dataset has been divided into three random subsets: 50% of the data was used for training, 25% was used for validation, and 25% was used for a test. All data categories (numerical, categorical, and ordinal) can be used as explanatory variables. The final matrix incorporates 6,853 rows with 79 attribute levels. The deep learning will consider training data to develop the learning framework.
- This study developed an initial model (Model 1) by using training data with one pass (known as epoch) over the training data. For Model 1, the size of hidden layer is 2.
- For Model 2, stopping criteria was applied. A larger number of epochs (100,000) was used to refine the model. For Model 2, the size of hidden layer is 32. It was found that the precision accuracy was not improving after 100 epochs.
- For the final model (Model 3), the deep learning algorithm was tuned with an adaptive learning algorithm. Two tuning parameters (ρ *and* ε) balance the global and local search efficiencies. The parameter ρ is the similarity to prior weight updates, and ε is a parameter that makes optimization work beyond local optima. Adaptive learning rate algorithm uses stochastic descent optimization. For the final model, these parameters were used:
  - Epochs: 100
  - Hidden layers: 128
  - Early stopping: enabled
  - Annealing rate: 2×10^{-6}
  - Samples = 350,000

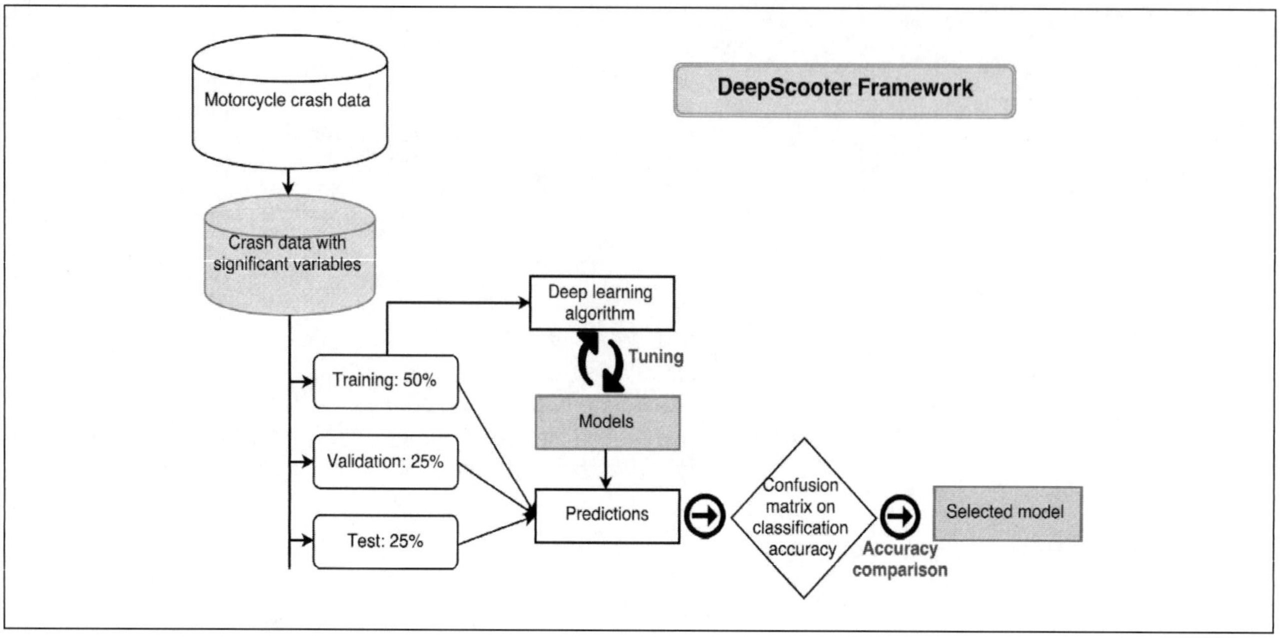

Figure 2. Framework of *DeepScooter* tool.

- Once the classifier has been trained (i.e., the parameters of the different layers of the model have been fixed), the quality of the classification outputs predicted by the model are compared against the correct "true" values stored in a labeled dataset. The confusion matrix describes the prediction accuracies and misclassification (error) rate. The outputs of the confusion matrix are reported in Tables 2 and 3.

Table 2 summarizes the results of Model 1 in confusion matrix format. For training data, the overall accuracies are 92.5%. Higher inaccuracies are found in identifying non-incapacitating injuries. For the validation dataset, the accuracy is lower than training data (around 90%). Both "*no injury*" and "*non-incapacitating injury*" show higher misclassification rates.

Table 2 also lists the results of Model 2 in confusion matrix format. For training data, the overall accuracies are improved by 3.5% (Model 1: 92.5% vs. Model 2: 96%). "*No injury*" showed a higher misclassification rate (around 8%). For the validation dataset, the accuracy was lower than training data (around 94%). "*No injury*" showed a higher misclassification rate (around 12%). For Model 1 and Model 2, results of text data are not shown.

Table 3 summarizes the results of Model 3 (training, validation, and test) in confusion matrix format. The findings include:

- For training data, the estimation accuracy of the training data was nearly 100%. Out of a sample size of 3440, the model can accurately classify 3437 crash severities.
- For validation set, the estimation accuracy was not improved from the accuracy derived in Model 2. "*Non-incapacitating injury*" showed higher misclassification.
- For the test set, the misclassification rates are 8% and 7% for Model 1, and Model 2, respectively. Model 3 results showed the lowest misclassification rate for test data. "*No injury*" showed higher misclassification for text data.

Figure 3*a* shows the overall misclassification rates of three models used for three sets of data. The training set showed that misclassification rate decreases sharply from 7% to 0%. For the validation set, the decrement is slower (from 9% to 6%). Test data also observed a slower decrease (from 8% to 6%). Overall, Model 3 performed better than the other two models. However, the validation set showed slightly a lower misclassification rate (5.79%) in Model 2 than the misclassification rate of Model 3 (6.25%).

Figure 3, *b–e*, illustrates comparison plots (Model 2 vs. Model 3) of the classification error and root mean square error (RMSE) over all epochs and samples. The values closer to zero indicate better model fit.

This study has applied two methods, multinomial logistic regression (statistical model) and support vector machine algorithm (machine learning method), on the same dataset. The highest prediction accuracy was found

Table 2. Confusion Matrix for Model 1 and Model 2

Severity level	Fatal	Incapacitating injury	Non-incapacitating injury	Possible injury	No injury	Column total	Error	Error rate	Metrics
Training (Model 1)									
Fatal	157	0	0	0	4	161	4	2.48	MSE: 0.0625
Incapacitating injury	1	245	0	0	2	248	3	1.21	RMSE: 0.250
Non-incapacitating injury	6	8	963	13	140	1130	167	14.78	logloss: 0.250
Possible injury	2	3	16	1129	9	1159	30	2.59	Mean per-class
No injury	1	2	9	41	689	742	53	7.14	Error: 0.056
Total	167	258	988	1183	844	3440	257	7.47	
Validation (Model 1)									
Fatal	69	0	0	0	3	72	3	4.17	MSE: 0.0795
Incapacitating injury	1	122	0	0	3	126	4	3.17	RMSE: 0.282
Non-incapacitating injury	4	6	520	7	75	612	92	15.03	logloss: 0.344
Possible injury	0	5	14	525	2	546	21	3.85	Mean per-class
No injury	0	3	10	27	315	355	40	11.27	Error: 0.075
Total	74	136	544	559	398	1711	160	9.35	
Training (Model 2)									
Fatal	161	0	0	0	0	161	0	0	MSE: 0.035
Incapacitating injury	1	246	0	0	1	248	2	0.81	RMSE: 0.189
Non-incapacitating injury	5	7	1085	7	26	1130	45	3.98	logloss: 0.146
Possible injury	3	4	21	1125	6	1159	34	2.93	Mean per-class
No injury	1	2	14	39	686	742	56	7.55	Error: 0.031
Total	171	259	1120	1171	719	3440	137	3.98	
Validation (Model 2)									
Fatal	71	0	1	0	0	72	1	1.39	MSE: 0.0532
Incapacitating injury	1	123	0	0	2	126	3	2.38	RMSE: 0.231
Non-incapacitating injury	4	6	584	5	13	612	28	4.58	logloss: 0.243
Possible injury	0	5	17	521	3	546	25	4.58	Mean per-class
No injury	0	2	14	26	313	355	42	11.83	Error: 0.049
Total	76	136	616	552	331	1711	99	5.79	

Table 3. Confusion Matrix for Model 3

Severity level	Fatal	Incapacitating injury	Non-incapacitating injury	Possible injury	No injury	Column total	Error	Error rate	Metrics
Training									
Fatal	161	0	0	0	0	161	0	0.00	MSE: 0.000
Incapacitating injury	0	248	0	0	0	248	0	0.00	RMSE: 0.025
Non-incapacitating injury	0	0	1128	0	2	1130	2	0.18	logloss: 0.003
Possible injury	0	0	0	1158	1	1159	1	0.09	Mean per-class
No injury	0	0	0	0	742	742	0	0.00	Error: 0.000
Total	161	248	1128	1158	745	3440	3	0.09	
Validation									
Fatal	70	1	0	0	1	72	2	2.78	MSE: 0.058
Incapacitating injury	1	124	1	0	0	126	2	1.59	RMSE: 0.241
Non-incapacitating injury	0	2	315	13	25	355	30	8.45	logloss: 0.544
Possible injury	4	6	13	582	7	612	33	5.39	Mean per-class
No injury	1	5	15	12	513	546	40	7.33	Error: 0.053
Total	76	138	344	607	546	1711	107	6.25	
Test									
Fatal	79	0	3	1	0	83	4	4.82	MSE: 0.058
Incapacitating injury	0	99	2	1	2	104	5	4.81	RMSE: 0.241
Non-incapacitating injury	3	4	511	8	12	538	27	5.02	logloss: 0.515
Possible injury	1	1	19	583	10	614	31	5.05	Mean per-class
No injury	1	3	9	29	321	363	42	11.57	Error: 0.063
Total	84	107	544	622	345	1702	109	6.40	

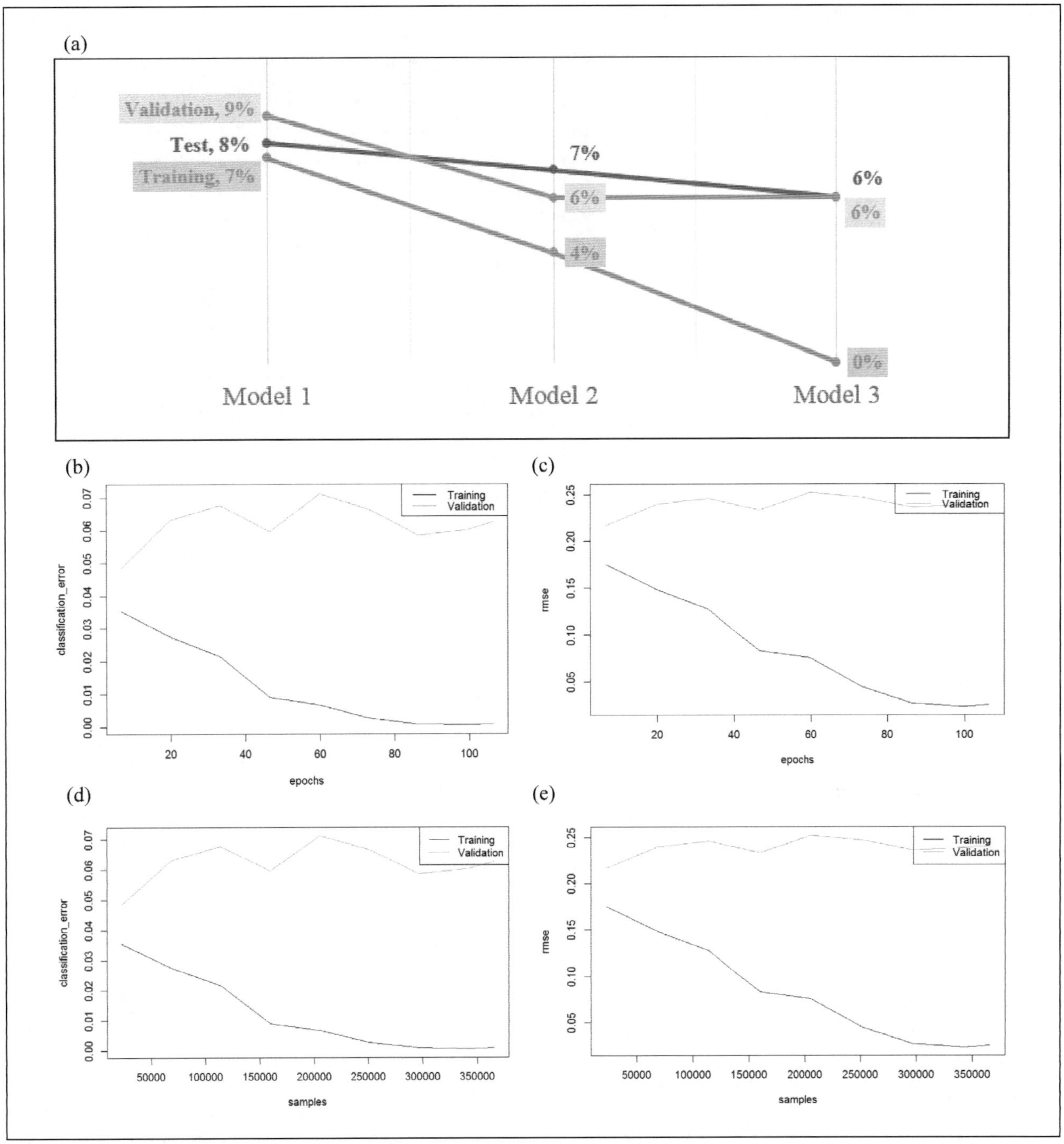

Figure 3. (*a*) Misclassification (error) rate in three models; (*b*) Model 2 classification error; (*c*) Model 2 RMSE; (*d*) Model 3 classification error; and (e) Model 3 RMSE.

as 78%, which is far below the attained accuracy from the deep learning method. The main contribution of this paper is the development of a deep learning framework to perform severity analysis of motorcycle crashes. Although the present study is focused on the improvement of the classification accuracy, a sound understanding of the explanatory variables would be beneficial. The heat chart shows (see Table 4) the conditional probability of significant variables from final deep learning model (Model 3). The color intensity (from red to white) varies along each

Table 4. Conditional Probabilities of Top Ten Attributes

Rider ejected	89.9	84.3	67.2	48.4	16.5
Two-way with no physical separation	62.3	62.3	63.2	61.6	60.4
State highway	48.4	29.7	38.7	43.9	38.1
Single vehicle	42.1	32.8	45.5	39.4	32.2
Two-way with physical separation	25.6	24.1	24.3	24	24.7
Business and mixed residential	24.1	28.2	26.4	27.7	35.5
Residential scattered	23.1	27.4	26.4	27.2	27.2
Saturday	22.5	21.3	18.4	20.9	18.8
Curve alignment	20.6	17.4	17.6	16.2	9.7
Sunday	19.9	19	17.6	14.5	15.2
	K	A	B	C	O

row. For example, around 90% of cases in fatal motorcycle crashes involve rider ejection. Fatal crashes are highly associated with rider ejection, a two-way road with no physical separation, single vehicle, curve aligned roadways, and weekends. Residential areas were not highly skewed in crash fatalities; this may be because of lower posted speeds in residential neighborhoods. Two-way roadways (both divided and undivided) are overrepresentative in severe and fatal crashes.

Conclusion

Many studies on motorcycle crash data have been conducted to understand the contributing factors that influence the severity of crashes. In 2014, the U.S. experienced 976 motorcycle crash-related fatalities. This statistics was more than twice the number of motorcycle rider fatalities that occurred in 1997. This unacceptably high number of motorcycle crashes calls for research to be conducted with additional resources and in newer directions. The 2010–2014 at-fault motorcycle rider crash data for Louisiana were used in this study. The findings include:

- The descriptive statistics show that motorcycle crashes generally feature high proportions in weekends, two-lane rural roads with no physical barriers, rider ejection, daylight, and clear weather conditions.
- The study confirms that, in modeling crash severity, the developed deep learning framework *DeepScooter* can estimate accurately up to 100%. The accuracy rate for test data ranges from 92% to 94%. The framework has sufficient reproducibility for use with a larger set of motorcycle crash data. For example, it can work as a suitable framework for identifying significant factors from FHWA Motorcycle Crash Causation Study (MCCS) (*49*).
- The conditional probability chart from *DeepScooter* shows that fatal crashes are more likely to be associated with rider ejection, a two-way road with no physical separation, single vehicle, curve aligned roadways, and weekends. Residential areas were not highly skewed in fatal crashes, as a result of lower posted speeds. In addition, younger riders are the vulnerable group in fatal motorcycle crashes.

The advantage of using a deep learning tool is its efficiency and high prediction accuracy. As this method does not require holding any statistical assumption, there is no consequence of biased results because of the violation of assumptions. However, this method has several limitations. One limitation of deep learning models is their low explanatory command because of the black box approach. In transportation safety research, interpretability is considered as one of the major constraints in adapting sophisticated deep learning models in real life. The current developed framework has the flexibility of reproduction that can be used by other researchers. It is anticipated that the *DeepScooter* framework and the findings will provide noteworthy contributions to the reduction of motorcycle crashes and crash-involved severities.

Author Contributions

The authors confirm contribution to the paper as follows: study conception and design: Subasish Das, Anandi Dutta; data collection: Subasish Das; analysis and interpretation of results: Subasish Das, Anandi Dutta; draft manuscript preparation: Subasish Das, Anandi Dutta, Lisa Minjares-Kyle, Karen Dixon, George Gillette. All authors reviewed the results and approved the final version of the manuscript.

References

1. National Highway Traffic Safety Administration. *Motorcycles.* https://www.nhtsa.gov/road-safety/motorcycles. Accessed July 3, 2017.
2. Louisiana Crash Data Reports. http://datareports.lsu.edu/reports.aspx?p = ci&yr = 2016&rpt = H1. Accessed August 1, 2017.
3. Lord, D., and F. Mannering. The Statistical Analysis of Crash-Frequency Data: A Review and Assessment of Methodological Alternatives. *Transportation Research Part A: Policy and Practice*, Vol. 44, No. 5, 2010, pp. 291–305.
4. Savolainen, P., F. Mannering, D. Lord, and M. Quddus. The Statistical Analysis of Highway Crash-Injury Severities: A Review and Assessment of Methodological Alternatives. *Accident Analysis & Prevention*, Vol. 43, No. 5, 2011, pp. 1666–1676.
5. Mannering, F., and C. Bhat. Analytic Methods in Accident Research: Methodological Frontier and Future Directions. *Analytic Methods in Accident Research*, Vol. 1, 2014, pp. 1–22.
6. Mannering, F., V. Sarkar, and C. Bhat. Unobserved Heterogeneity and the Statistical Analysis of Highway Accident Data. *Analytic Methods in Accident Research*, Vol. 11, 2016, pp. 1–16.
7. Savolainen, P., and F. Mannering. Probalistic Models of Motorcyclists' Injury Severities in Single-and Multi-Vehicle Crashes. *Accident Analysis and Prevention*, Vol. 39, 2007, pp. 955–963.
8. Lapparent, M. Empirical Bayesian Analysis of Accident Severity for Motorcyclists in Large French Urban Areas. *Accident Analysis and Prevention*, Vol. 38, 2006, pp. 260–268.
9. Schneider, W. H., P. T. Savolainen, D. V. Boxel, and R. Beverly. Examination of Factors Determining Fault in Two-Vehicle Motorcycle Crashes. *Accident Analysis and Prevention*, Vol. 45, 2012, pp. 669–676.
10. Turner, P., L. Higgins, and S. Geedipally. *Development of Statewide Motorcycle Safety Plan for Texas.* Technical Report, FHWA/TX-12-0-6712-1, 2013.
11. Quddus, M. A., R. B. Noland, and H. C. Chin. An Analysis of Motorcycle Injury and Vehicle Damage Severity Using Ordered Probit Models. *Journal of Safety Research*, Vol. 33, 2002, pp. 445–462.
12. Cunto, F. J. C., and S. Ferreira. An Analysis of the Injury Severity of Motorcycle Crashes in Brazil Using Mixed Ordered Response Models. *Journal of Transportation Safety & Security*, Vol. 9, No. S1, 2017, pp. 33–46.
13. Shaheed, M., and K. Gkritza. A Latent Class Analysis of Single-Vehicle Motorcycle Crash Severity Outcomes. *Analytic Methods in Accident Research*, Vol. 2, 2014, pp. 30–38.
14. Shaheed, M., K. Gkritza, W. Zhang, and Z. Hans. A Mixed Logit Analysis of Two-Vehicle Crash Severities Involving a Motorcycle. *Accident Analysis and Prevention*, Vol. 61, 2013, pp. 119–128.
15. Shankar, V., and F. Mannering. An Exploratory Multinomial Logit Analysis of Single-Vehicle Motorcycle Accident Severity. *Journal of Safety Research*, Vol. 27, No. 3, 1996, pp. 183–194.
16. Geedipally, S. R., P. T. Turner, and S. Patil. Analysis of Motorcycle Crashes in Texas with Multinomial Logit Model. *Transportation Research Record: Journal of the Transportation Research Board*, 2011. 2265: 62–69.
17. Chung, Y., T. J. Song, and B. J. Yoon. Injury Severity in Delivery-Motorcycle to Vehicle Crashes in the Seoul Metropolitan Area. *Accident Analysis and Prevention*, Vol. 62, 2014, pp. 79–86.
18. Kim, K., J. Boski, and E. Yamshita. Typology of Motorcycle Crashes: Rider Characteristics, Environmental Factors and Spatial Patterns. *Transportation Research Record: Journal of the Transportation Research Board*, 2002. 1818: 47–53.
19. Schneider, W. H., and P. T. Savolainen. Comparison of Severity of Motorcyclist Injury by Crash Types. *Transportation Research Record: Journal of the Transportation Research Board*, 2011. 2265: 70–80.
20. Blackman, R. A., and N. L. Haworth. Comparison of Moped, Scooter and Motorcycle Crash Risk and Crash Severity. *Accident Analysis and Prevention*, Vol. 57, 2013, pp. 1–9.
21. Pai, C., and W. Saleh. Exploring Motorcyclist Injury Severity Resulting from Various Crash Configurations at T-Junctions in the United Kingdom: An Application of the Ordered Probit Models. *Traffic Injury Prevention*, Vol. 8, 2007, pp. 62–68.
22. Rifaat, S. M., R. Tay, and A. de Barros. Severity of Motorcycle Crashes in Calgary. *Accident Analysis and Prevention*, Vol. 49, 2012, pp. 44–49.
23. Savolainen, P. T., F. L. Mannering, D. Lord, and M. A. Quddus. The Statistical Analysis of Highway Crash-Injury Severities: A Review and Assessment of Methodological Alternatives. *Accident Analysis and Prevention*, Vol. 43, 2011, pp. 1666–1676.
24. Haque, M., H. C. Chin, and A. K. Debnath. An Investigation on Multi-Vehicle Motorcycle Crashes Using Log-Linear Models. *Safety Science*, Vol. 50, 2012, pp. 352–362.

25. Allen, T., S. Newstead, M. G. Lenne, R. McClure, P. Hillard, M. Symmons, and L. Day. Contributing Factors to Motorcycle Injury Crashes in Victoria, Australia. *Transportation Research Part F: Traffic Psychology and Behavior*, Vol. 45, 2017, pp. 157–168.
26. Chen, C., G. Zhang, X. Liu, Y. Ci, H. Huang, J. Ma, Y. Chen, and H. Guan. Driver Injury Severity Outcome Analysis in Rural Interstate Highway Crashes: A Two-Level Bayesian Logistic Regression Interpretation. *Accident Analysis and Prevention*, Vol. 97, 2016, pp. 69–78.
27. Breiman, L. Statistical Modeling: The Two Cultures (With Comments and a Rejoinder by the Author). *Statistical Science*, Vol. 16, No. 3, 2001, pp. 199–231.
28. Das, S., and X. Sun. Association Knowledge for Fatal Run-off-Road Crashes by Multiple Correspondence Analysis. *IATSS Research*, Vol. 39, No. 2, 2016, pp. 146–155.
29. Chen, C., and Y. Xie. Machine Learning for Recognizing Driving Patterns of Drivers of Large Commercial Trucks. *Transportation Research Record: Journal of the Transportation Research Board*, 2015. 2517: 18–27.
30. Das, S., A. Dutta, R. Avelar, K. Dixon, X. Sun, and M. Jalayer. Supervised Association Rules Mining on Pedestrian Crashes in Urban Areas: Identifying Patterns for Appropriate Countermeasures. *International Journal of Urban Sciences*, 2018. https://doi.org/10.1080/12265934.2018.1431146
31. Sun, J., J. Sun, and P. Chen. Use of Support Vector Machine Models for Real-Time Prediction of Crash Risk on Urban Expressways. *Transportation Research Record: Journal of the Transportation Research Board*, 2014. 2432: 91–98.
32. Iranitalab, A., and A. Khattak. Comparison of Four Statistical and Machine Learning Methods for Crash Severity Prediction. *Accident Analysis and Prevention*, Vol. 108, 2017, pp. 27–36.
33. Das, S., L. Minjares-Kyle, R. Avelar, K. Dixon, and B. Bommanayakanahalli. Improper Passing-Related Crashes on Rural Roadways: Using Association Rules Negative Binomial Miner. Presented at 96th Annual Meeting of the Transportation Research Board, Washington, D.C., 2017.
34. Dong, N., H. Huang, and L. Zheng. Support Vector Machine in Crash Prediction at the Level of Traffic Analysis Zones: Assessing the Spatial Proximity Effects. *Accident Analysis & Prevention*, Vol. 82, 2015, pp. 192–198.
35. Das, S., and X. Sun. Factor Association with Multiple Correspondence Analysis in Vehicle-Pedestrian Crashes. *Transportation Research Record: Journal of the Transportation Research Board*, 2015. 2519: 95–103.
36. Ghazan Khan, G., A. Bill, and D. Noyce. Exploring the Feasibility of Classification Trees Versus Ordinal Discrete Choice Models for Analyzing Crash Severity. *Transportation Research Part C: Emerging Technologies*, Vol. 50, 2015, pp. 86–96.
37. Das, S., R. Avelar, K. Dixon, and X. Sun. Investigation on the Wrong Way Driving Crash Patterns Using Multiple Correspondence Analysis. *Accident Analysis and Prevention*, Vol. 111, 2018, pp. 43–55.
38. Mishra, D., eds. Information and Communication Technology. *Advances in Intelligent Systems and Computing*, Vol. 625, 2018.
39. Das, S., A. Mudgal, A. Dutta, and S. Geedipally. Vehicle Consumer Complaint Reports Involving Severe Incidents: Mining Large Contingency Tables. *Transportation Research Record: Journal of the Transportation Research Board*, 2018.
40. Das, S., A. Dutta, M. Jalayer, A. Bibeka, and L. Wu. Factors Influencing the Patterns of Wrong Way Driving Crashes on Freeway Exit Ramps and Median Crossovers: Exploration Using 'Eclat' Association Rules to Promote Safety. *International Journal of Transportation Science and Technology*, 2018.
41. Duan, Y., Y. Lv, Y. L. Liu, and F. Y. Wang. An Efficient Realization of Deep Learning for Traffic Data Imputation. *Transportation Research Part C: Emerging Technologies*, Vol. 72, 2016, pp. 168–181.
42. Polson, N. G., and V. O. Sokolov. Deep Learning for Short-Term Traffic Flow Prediction. *Transportation Research Part C: Emerging Technologies*, Vol. 79, 2017, pp. 1–17.
43. Yu, S., Y. Wu, W. Li, Z. Song, and W. Zeng. A Model for Fine-Grained Vehicle Classification Based on Deep Learning. *Neurocomputing*, Vol. 257, 2017, pp. 97–103.
44. Rosenblatt, F. *The Perceptron: A Perceiving and Recognizing Automaton*. Report No. 85-460-1. Cornell Aeronautical Laboratory, Ithaca, NY, 1957.
45. Zhong, G., L. Wang, X. Ling, and J. Dong. An Overview on Data Representation Learning: From Traditional Feature Learning to Recent Deep Learning. *The Journal of Finance and Data Science*, Vol. 2, 2016, pp. 265–278.
46. Le, Q. A Tutorial on Deep Learning Part 2: Autoencoders, Convolution Neural Networks and Recurrent Neural Networks. http://robotics.standford.edu/~quocle/tutorial2.pdf. Accessed July 3, 2017.
47. Goodfellow, I., Y. Bengio, and A. Courville. Deep Learning. The MIT Press, Cambridge, MA, 2016.
48. H2O.ai Team. *H_2o: R Interface for H2O*. R Package Version 3.1.0, 2015.
49. Nazemetz, J., F. Bents, J. Perry, C. Thor, and C. Tan. *Motorcycle Crash Causation Study: Final Report*. FHWA, 2016.

The Standing Committee on Motorcycles and Mopeds (ANF30) peer-reviewed this paper (18-04840).